# ANEMONE FISHES

### AND THEIR HOST SEA ANEMONES

# ANEMONE FISHES

## AND THEIR HOST SEA ANEMONES

A Guide for Aquarists and Divers

Daphne G Fautin
and Gerald R Allen

Western Australian Museum

© Western Australian Museum 1992
Revised Edition 1997. ISBN 0 7309 8365 X
Design: Howard Osborne. Cover design: Robyn Mundy
Printed by Kaleidoscope, West Perth WA 6005
Published by Western Australian Museum,
Francis Street, Perth, WA 6000

Frontispiece: *Amphiprion bicinctus* with *Heteractis crispa* in the Red Sea. (G. Douwmal/Ikan)

# Foreword

Anemonefishes and their invertebrate hosts have delighted the western world since 1881 when the first captive specimens were kept in a tub of seawater. However, it was not until the mid-20th century that the intimate relationship of these tropical animals began to be known worldwide. With the advent of SCUBA diving and the establishment of commercial air routes to equatorial destinations in the Indian and Pacific Oceans, pristine coral reefs became accessible to an increasing audience. Skin-diving tourists, sport divers, naturalists, and marine scientists have all helped contribute to underwater discoveries, among them the fascinating natural history of anemonefishes. Virtually all large public aquaria have at least one anemonefish display, and these animals have been at or near the top in aquarium fish sales for the past three decades, attesting to their tremendous popularity.

In view of this unprecedented public exposure to the fascinating relationship between sea anemones and fishes, we have written a book with needs at all levels, from teen-age aquarist to research scientist, in mind. Because of confusion in sea anemone taxonomy, previous works on this subject often used incorrect or outdated names. This book permits quick and accurate identification of the invertebrate hosts, as well as the fishes, through well-illustrated, easy-to-use keys and underwater photographs. It is the first publication on these animals designed as a field guide. We hope that it will add even greater pleasure to your fish-watching endeavours and provide new insights into the symbiotic relationship of fishes and sea anemones.

We are grateful to the Christensen Research Institute for making our long-held dream of writing this book come true, and we dedicate this book to the memory of Reed Fautin, who taught us both ecology.

Daphne G. Fautin<br>
Gerald R. Allen

*Amphiprion perideraion* with *Heteractis crispa*, Philippine Islands. (H. Hall/Ikan)

# Contents

|  | PAGE |
|---|---|
| FOREWORD | v |
| INTRODUCTION | 1 |
|   Geographical and ecological distribution | 1 |
|   How is this relationship possible? | 4 |
|   Other symbionts | 7 |
|   Scientific names and what they mean | 7 |
|   How to use this guide | 11 |
| CHAPTER 1: SEA ANEMONES | 13 |
|   Classification | 13 |
|   Identification | 14 |
|   Identification aids | 18 |
|     Anatomical key | 19 |
|     Habitat key | 20 |
|   *Cryptodendrum adhaesivum* | 24 |
|   *Entacmaea quadricolor* | 26 |
|   *Heteractis aurora* | 28 |
|   *Heteractis crispa* | 30 |
|   *Heteractis magnifca* | 32 |
|   *Heteractis malu* | 34 |
|   *Macrodactyla doreensis* | 36 |
|   *Stichodactyla gigantea* | 38 |
|   *Stichodactyla haddoni* | 40 |
|   *Stichodactyla mertensii* | 42 |
| CHAPTER 2: ANEMONEFISHES | 45 |
|   Classification | 45 |
|   Identification | 46 |
|   Colour variation | 49 |
|   Key to anemonefishes | 52 |
|   *Amphiprion akallopisos* | 60 |
|   *Amphiprion akindynos* | 62 |
|   *Amphiprion allardi* | 64 |
|   *Amphiprion bicinctus* | 66 |
|   *Amphiprion chagosensis* | 68 |
|   *Amphiprion chrysogaster* | 70 |
|   *Amphiprion chrysopterus* | 72 |
|   *Amphiprion ciarkii* | 74 |
|   *Amphiprion ephippium* | 76 |
|   *Amphiprion frenatus* | 78 |
|   *Amphiprion fuscocaudatus* | 80 |
|   *Amphiprion latezonatus* | 82 |

# CONTENTS

## CONTENTS

| | PAGE |
|---|---|
| *Amphiprion latifasciatus* | 84 |
| *Amphiprion leucokranos* | 86 |
| *Amphiprion mccullochi* | 88 |
| *Amphiprion melanopus* | 90 |
| *Amphiprion nigripes* | 92 |
| *Amphiprion ocellaris* | 94 |
| *Amphiprion omanensis* | 96 |
| *Amphiprion percula* | 98 |
| *Amphiprion perideraion* | 100 |
| *Amphiprion polymnus* | 102 |
| *Amphiprion rubrocinctus* | 104 |
| *Amphiprion sandaracinos* | 106 |
| *Amphiprion sebae* | 108 |
| *Amphiprion thiellei* | 110 |
| *Amphiprion tricinctus* | 112 |
| *Premnas biaculeatus* | 114 |
| *Dascyllus trimaculatus* | 116 |
| CHAPTER 3: BIOLOGY OF SEA ANEMONES | 119 |
| Nutrition | 119 |
| Survival | 120 |
| Reproduction | 120 |
| Locomotion | 122 |
| Anemone-like animals | 123 |
| CHAPTER 4: LIFE HISTORY OF ANEMONEFISHES | 125 |
| Courtship, spawning, and egg-care | 126 |
| Larval life and settlement | 130 |
| Social structure and sex reversal | 131 |
| Feeding and growth | 132 |
| Hybrid Anemonefishes | 133 |
| CHAPTER 5: INTERACTIONS BETWEEN FISH AND SEA ANEMONES | 135 |
| Symbiosis | 136 |
| Specificity | 138 |
| How anemone and fish can affect one another | 140 |
| CHAPTER 6: AQUARIUM CARE | 143 |
| APPENDIX | 149 |
| ACKNOWLEDGEMENTS | 150 |
| REFERENCES AND RECOMMENDED READING | 151 |
| GLOSSARY | 155 |
| INDEX | 157 |

# Introduction

> While standing in the water, breast high, admiring this splendid zoophyte [sea anemone], I noticed a very pretty little fish which hovered in the water close by, and nearly over the anemone. This fish was six inches long, the head bright orange, and the body vertically banded with broad rings of opaque white and orange alternatively, three bands of each . . . . I made several attempts to catch it; but it always eluded my efforts —not darting away, however, as might be expected, but always returning presently to the same spot. Wandering about in search of shells and animals, I visited from time to time the place where the anemone was fixed, and each time, in spite of all my disturbance of it, I found the little fish there also.
>
> — Dr. Cuthbert Collingwood: *Rambles of a Naturalist on the Shores and Waters of the China Sea* (John Murray, London, 1868, page 151)

That was how the first person to record the remarkable living arrangement of anemonefishes (or clownfishes) and some sea anemones described his discovery on Fiery Cross Reef, off the shores of Borneo. He was captivated by what continues to fascinate anyone who has had the good fortune to see these animals "at home," in the shallow tropical waters of the Indian and Pacific Oceans. It is their beauty, and it is their intimate symbiosis. (Symbiosis, a word that literally means "living together," is used by scientists to describe the relationship between unrelated species of plants and/or animals that live in close association.)

The symbiosis between clownfishes and sea anemones fascinates diver and biologist alike, for many of the same reasons. The sea anemones that play host to the fishes are, like most of their kin, virtually immovably fixed. Each anemone constitutes the territory of its fish, which therefore seldom venture far from it, retreating into its tentacles when feeling threatened. This sedentariness, which so intrigued Dr. Collingwood, allows biologists to do long-term studies, revisiting the same animals repeatedly. Underwater enthusiasts, having once found them, can be assured of relocating fish and anemones on future dives. The fish can be approached *very* closely, and both partners are extremely beautiful, making them a prime subject of photographers.

## GEOGRAPHICAL AND ECOLOGICAL DISTRIBUTION

Sea anemones live throughout the world's oceans, from poles to equator, and from the deepest trenches to the shores, as do fishes. But no one kind of either lives in all places. Of nearly 1000 species of sea anemones, only 10 are host to anemonefishes. They live in the parts of the Indian and Pacific Oceans that lie within the tropics or where warm, tropical waters

*Amphiprion bicinctus* with *Entacmaea quadricolor* in the Red Sea. (C. Newbert/Ikan)

# INTRODUCTION

A pair of *Amphiprion clarkii* resting snugly for the night among the tentacles and oral disc folds of *Stichodactyla mertensii*. (E. Robinson/Ikan)

Anemonefishes exist no deeper than diving depths. *Amphiprion perideraion* with *Heteractis magnifica* on the mast of shipwreck in Chuuk Lagoon, central Pacific. (H. Hall/Ikan)

are carried by currents, such as the east coast of Japan (as far north as the latitude of Tokyo!). Because the 28 species of clownfishes live only with these 10 species of sea anemones, they are found in the same places.

The richest part of the world for these animals is around New Guinea: we have found eight species of fishes and anemones in the d'Entrecasteaux Islands off the east coast of that island, and all 10 host anemones with nine fishes around Madang, on the north coast of Papua New Guinea. The numbers of both diminish outward from there. Typically, a central Indo-West Pacific locality such as Guam, or Lizard Island on the Great Barrier Reef, has up to five different species of fishes and about an equal number of anemone species. Numbers are even smaller at the peripheries of their range. For example, one kind of fish and three of host anemones are known from the Comoro Islands, and no clownfishes but one host sea anemone species occurs in Hawaii.

These anemones and their anemonefishes exist only in shallow water, no deeper than SCUBA-diving depths. That is because within the cells of an anemone's tentacles and oral disc live microscopic, single-celled, golden-brown algae (dinoflagellates) called zooxanthellae. Like all plants, they require sunlight for photosynthesis, a process in which solar energy is used to make sugars from carbon and water. Some of these sugars fuel the algae's metabolism, but most of them "leak" to the anemone, providing energy to it. Therefore, the anemones that are host to fishes must live in sunny places. The amount of light in the sea diminishes rapidly with depth because water filters out sunlight. Turbidity also diminishes light penetration. So these anemones live at depths of no more than about 50 m, generally in clear water. (Reef-forming corals also contain algae, and coral reefs occur only in shallow, mostly clear water for the identical reason.)

Anemonefishes live in habitats other than reefs, but are usually thought of as reef dwellers because that is where most tropical diving occurs. Other habitats may be less colourful and diverse than reefs, but they can be equally fascinating. About as many species of host actinians

# INTRODUCTION

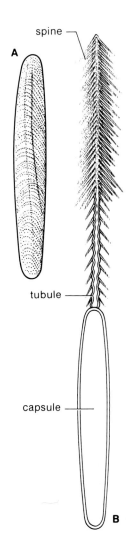

Structure of a nematocyst:
A - undischarged;
B - discharged.

(= sea anemones) live on sand-flats surrounding coral reefs, or even at some distance from reefs, as live on reefs themselves. Individuals of some species can survive in muddy areas, but they generally lack fish symbionts. Even on reefs, most species of host actinians are inconspicuous, unlike their partner fish. Spotting the fish first, then frightening it so that it takes refuge in its anemone, or (preferably) waiting patiently for its periodic bathe among the tentacles, is often the best way to locate an actinian.

## *HOW IS THIS RELATIONSHIP POSSIBLE?*

At the time of Collingwood's discovery, some species of fishes and anemones involved in this relationship had been known to science for a century already. Why had nobody reported their living arrangement before? We can only speculate. Perhaps poisons had been used to collect the fish, which causes them to float to the surface, so nobody knew where they had come from. Perhaps collectors saw fish living in anemones but did not appreciate its significance. Or quite possibly it was seen and simply not believed, so unlikely is an anemone as home to a fish.

Lovely, accessible — and a most unlikely partnership. Sea anemones are related to corals and more distantly to jellyfishes. Common to all of these animals are nematocysts, the harpoon-like stinging capsules that give jellyfish their sting, fire coral their burn, and the tentacles of some sea anemones their stickiness. The microscopic nematocysts, which are manufactured inside cells (but are not themselves cells), are particularly dense in tentacles and internal structures. Those of the tentacles function in defence and prey capture; internal nematocysts are essential to digestion. Within each capsule is coiled a fine tubule many times the capsule's length. When the capsule is stimulated to fire (a combination of chemical and mechanical stimuli is necessary to trigger most kinds of nematocysts; there are over 30 in all), the tubule shoots out, everting like the sleeve of a coat turned inside out, to penetrate or wrap around the target. Many types of nematocysts, although probably not all, contain toxins, which are delivered to predator and prey by or through the everting tubule.

The existence and function of nematocysts were known before the anemonefish symbiosis was described. And so, when Collingwood first reported "the discovery of some Actiniae of enormous size, and of habits no less novel than striking," his prime concern was with how the fish managed to survive in an environment that is deadly to most fishes, even some much larger than anemonefishes.

Over the years, many biologists have suggested ways in which it might be possible for the fish to survive in its hostile environment. Among the hypotheses [and reasons for discarding them] were the following. 1) Tentacles of these particular anemones do not contain nematocysts. [Not only are there nematocysts, but those of all 10 species of host actinians are typical in kind and quantity to those occurring in the majority of sea

# INTRODUCTION

anemones.] 2) The fish do not actually touch the tentacles. [While this is certainly true of some Caribbean fishes that seek protection behind and under sea anemones, genuine anemonefishes swim among tentacles, and sleep on the oral disc at night.] 3) The skin of anemonefishes is thicker than normal so nematocysts cannot penetrate it. [It differs little from that of other damselfishes, and may even be slightly thinner. Indeed, an unprotected anemonefish can be killed by its host's sting.] 4) While a fish is present, the anemone will not fire its nematocysts. [Although a sea anemone can exert some control over firing, this cannot be the solution to the riddle, because an actinian can sting and capture prey while harbouring clownfish.]

The young of the wrasse *Thalassoma amblycephalus* is sometimes seen with anemones (in this case *Heteractis magnifica*), but make only brief contact and are not dependent on the anemones. (R. Eisenhart)

Anemonefishes are easily kept in aquaria, many of which are as large as the fish's normal territory. Both fishes and sea anemones survive, apparently quite well, when separated from one another. However, if the separation lasts more than a few days or weeks, depending on the species involved, when the partners are reunited and the fish swims into the host's tentacles, it withdraws rapidly, appearing (sometimes very obviously, sometimes less so) to have been stung. Thus the protection of the fish is elicited or acquired, and can disappear. A fish that has been living alone will be stung by an anemone in which another anemonefish is being harboured. So the fish, rather than the actinian, is responsible for the protection.

But a stung anemonefish returns to its host repeatedly, going through an elaborate, stereotyped swimming dance, gingerly touching tentacles first to its ventral fins only, then to its entire belly. Finally, after a few minutes to several hours of such "acclimation" behaviour, it is able to dive right in.

Some clownfishes nibble at their host's tentacles, which it had been speculated might immunize them against the sting. But the fish are not

# INTRODUCTION

immune to being stung, as is sometimes stated. Immunity is a physiological response that extends throughout an animal's body. Experiments by Davenport and Norris conclusively proved that the protective agent resides in the mucus coating that anemonefishes, like all fishes, have on their surface. But what is the source of this protective mucus?

One theory is that it comes from the host actinian. Supporters of this theory (Schlichter foremost among them) believe that during its elaborate "acclimation" swimming, when contact is initially made with its host, the fish smears mucus from the anemone all over itself. Just as the sea anemone does not sting itself, it does not sting a fish, or any other object, covered in its mucus. The fish is thereby chemically camouflaged: it is, essentially, a fish in anemone's clothing. The fish's normal behaviour of returning to its anemone at least once a minute can be interpreted as serving to maintain its protective layer of mucus. According to this theory, what allows clownfishes to live in this peculiar habitat is their unusual behaviour.

Finding anemone mucus on many objects with which the animal regularly comes in contact, such as the rocks and algae around it, other scientists (Lubbock foremost among them) believe that its presence on a fish is the *result* of the fish's being protected rather than its *cause*. The fish's own mucus has evolved to lack components that stimulate nematocyst discharge, according to this theory, and "acclimation" behaviour may be an artifact of artificially separating animals that normally never are parted. The secret to clownfishes' peculiar habitat, according to this interpretation, is their unusual biochemistry.

As in so much of science, there is probably truth on both sides. Although all anemonefishes are closely related and share an unusual habitat, they vary in some aspects of their biology, including how far they venture from their home, how many fish occupy a single anemone, and which hosts and how many host species they occupy (Table 1). Similarly, they may not all adapt to an actinian in the identical manner, as is generally assumed, with behaviour and biochemistry probably both playing roles to varying degrees. We believe that for fish that live with many types of hosts (such as *Amphiprion clarkii*, which is the least host-specific), behaviour is likely to be more important to adaptation, whereas for host-specific fish (such as *Premnas biaculeatus*), biochemistry is probably the more significant factor. An experiment by Brooks and Mariscal provided evidence that both fish and anemone may be active in forming the symbiosis for at least one combination of fish and anemone species. The average acclimation time following prolonged separation of *A. clarkii* from the host anemone *Macrodactyla doreensis* was two and a half hours, but a fish kept in an aquarium with a surrogate sea anemone made of rubber bands glued to a Petri dish required an average of only 20 minutes to acclimate to a real anemone. Thus it would appear that the fish does produce an especially protective mucus when living in what it perceives to be a sea anemone, but since it must still

# INTRODUCTION

undergo a period of acclimation, that alone does not suffice. Presumably, the anemone alters what is there, or adds to it.

## OTHER SYMBIONTS

Some host actinians harbour small, mostly black damselfish of the genus *Dascyllus*. However, this fish is not considered to be a true anemonefish, as are the members of *Amphiprion* and *Premnas*, because their lives are not dependent on the anemone host. Rather, when small, they may live with actinians; as they grow, they become independent. Sometimes *Dascyllus* is the only fish in an anemone; sometimes it shares its host with true anemonefish. In addition, other fishes, particularly young wrasses, are sometimes seen in close proximity to sea anemones, although only members of *Thalassoma* make infrequent contact with the tentacles. Organisms that may, but need not, live with those of another, unrelated species are termed "facultative symbionts." Those that must, such as anemonefishes, are "obligate symbionts."

In addition to having algal symbionts and fish symbionts, some species of these sea anemones harbour shrimps and crabs. Because of their small size, because they cannot easily be tagged or recognized individually, and because they scurry under the overhang of the anemone's oral disc when disturbed, they have hardly been studied. Therefore, in contrast to clownfishes, we do not know how closely their lives are tied to the anemones, and whether they are obligately or facultatively symbiotic.

Some sea anemones outside the tropical Indo-West Pacific have facultative fish associates. No fewer than 30 species of Caribbean fishes may occur in or near anemones. Some, like *Dascyllus*, engage in this association only when small. Most never actually come into contact with the anemone's tentacles, but merely hover around them. Presumably, even proximity to an anemone confers some benefit. In the cold waters of British Columbia, Canada, some small individuals of the convict fish, *Oxylebias pictus*, sleep lying in the tentacles of the sea anemone *Urticina lofotensis*. They leave during the day, but return to the same anemone each night. As the fish grows, it returns less regularly, until it outgrows its "security blanket" altogether.

## SCIENTIFIC NAMES AND WHAT THEY MEAN

In this book, we use both common and scientific names for fishes and anemones. The common names are in English; a Japanese translation of our book would use an entirely different set of Japanese common names. But a scientific name is always in Latin. This began in the era when Latin was the language of learning, and assures that reader and writer, or speaker and listener need not share a language: regardless of what languages they speak, both mean the same kind of animal when they use a particular name.

A Latin, or scientific, name consists of two parts — a genus (always capitalised; its plural is genera) and a species (never capitalised; its plural

# INTRODUCTION

Young of *Dascyllus trimaculatus* with *Stichodactyla haddoni* at Madang, Papua New Guinea. The anemonefish is *Amphiprion polymnus*. (R. Steene)

Anemone crab with *Cryptodendrum adhaesivum*. (R. Steene)

Anemone shrimp among tentacles of *Heteractis magnifica*. (R. Steene)

# INTRODUCTION

is also species). It is italicized (in print) or underlined (in typing or hand-writing) to identify it as being in Latin. Similar species are placed in the same genus. Two species in the same genus are more alike, and are considered more closely related evolutionarily, than are two species in different genera. The first time a generic name is used in a piece of writing, it must be spelled out, but subsequently it may be abbreviated in a way that will not cause confusion. In this book, for example, *A.* stands for *Amphiprion*. A generic name may be used by itself, without a species name, but the reverse is not allowed. That is because the same species name may be used for species that belong to different genera — but each combination of the two is unique.

The generic and/or specific name of an organism often alludes to some conspicuous feature of it, or to where it lives. For example, *Stichodactyla gigantea* is among the largest sea anemones, and *Amphiprion chagosensis* lives only in the Chagos Archipelago of the Indian Ocean. It may also honour someone important in its discovery, such as *S. haddoni*, which memorialises Alfred C. Haddon, one of those 19th century all-round naturalists, who did anthropological research in northern Australia, but collected sea anemones on the side.

We provide the most accurate, current name for each species of fish and actinian, as well as some of the other names (synonyms) by which it has been known. Although each species has only one valid scientific name, it may have been given other names during the past. This is frequently the case for highly variable animals, such as *Amphiprion clarkii*, to which more than 10 names have been applied! In our studies of hundreds of animals from many localities, we were unable to find consistent, significant differences in those that have been called by different names, and so concluded that the name *A. clarkii* is appropriate for all of them. There is another reason for multiple names. Several species of host sea anemones were first found in the Red Sea. Animals of the same species subsequently collected from the East Indies or Australia were commonly given new names. This may have been partly because communications were poor in those days — scientists were often unaware that a particular species had already been described. But more likely, early biologists probably did not imagine that animals collected thousands of kilometers apart might belong to the same species. So which of the many names should be used? Usually, it is the oldest one. This gives the first person to have named it credit for the discovery. For example, although *Radianthus ritteri* is a name that has commonly been used for one species of sea anemone, the valid name for that species is the earlier described *Heteractis magnifica*; the former dates from 1898, and the latter from 1830.

Each binomial scientific name uniquely applies to only one species, but the following example illustrates how multiple uses for the same name may arise. *Stichodactyla gigantea* was named by Europeans to whom anemones of this species, first seen in the Red Sea, were indeed gigantic, in comparison with most temperate anemones. We now know

# INTRODUCTION

*Amphiprion percula* with rare colour variety of *Stichodactyla gigantea*, Madang, Papua New Guinea. (G. Allen)

that the name *S. gigantea* does not belong to the largest sea anemone, although that is a logical assumption, leading some people to apply that name to the largest, which is actually *S. mertensii*. It is especially important in symbioses to identify the partners accurately because the relationships between different partners are not necessarily the same.

The name(s) of the person(s) who described the species, and the year in which the description was published, are part of the formal scientific name, but are not always appended to it. If a particular name has been used more than once, this allows the reader to know in which sense it is being used and to determine when the name was proposed. If the name of the author is in parentheses, the species has been transferred from the genus in which it was originally described. For example, *Heteractis magnifica* was originally named *Actinia magnifica* by Quoy and Gaimard in 1830, when nearly all sea anemones were placed in the genus *Actinia*. As more species were discovered, and we learned which are closely and which distantly related, additional generic names were created, each referring to a group of closely related species.

The names *Stichodactyla* and *Heteractis* are used for anemones that were, in general, previously referred to as *Stoichactis* and *Radianthus*, respectively, in the technical literature, and still are in some popular writing. The names we use were selected not only because they are senior, but more importantly because the old names were used inconsistently. Both names, for example, had been used to refer to species in both genera. Species of other genera had also been referred to them, and species that belong in *Stichodactyla* and *Heteractis* had been called by other generic names. We hope that this book will set a standard so that divers, hobbyists, aquarists, and scientists alike will use the same name in the same way, avoiding such confusion in future.

# INTRODUCTION

## *HOW TO USE THIS GUIDE*

In the following two chapters, we explain how to identify the 10 species of host sea anemones and the 28 species of anemonefishes, provide keys to them, and characterise each. We describe the morphological features of an animal most important to its identification in the field, as well as geographical range and typical habitat.

The most important aspect of a fish's habitat is obviously its host; the anemone's habitat specificity will govern where fish associated with it are found. Species of partner may or may not be an important clue in identification. For example, if you find a fish you can identify as *P. biaculeatus*, you have also identified the anemone with which it is living, for it is known only from *Entacmaea quadricolor*. However, identifying an anemone as *E. quadricolor* is not much help in identifying a fish occupying it; 12 species of clownfishes other than *P. biaculeatus* are symbiotic with it. Conversely, finding *Amphiprion clarkii* in an anemone is no help, because it occurs with all 10 host actinians. But the fish in *Cryptodendrum adhaesivum* is certainly *A. clarkii*, the only fish known to live with it. Using a symbiont to identify its partner is valid only in the field. In captivity, many fish can adapt to living with almost any host actinian, in addition to numerous exotic ones (that is, anemones from elsewhere, with which anemonefishes are not naturally found).

Usually fish of only one species occur with an individual actinian, but occasionally fish of two species may share a host (usually a large one, which they divide into exclusive territories, so in a sense they are not actually sharing the space). It is possible that rarely three may coexist. But these instances are sufficiently uncommon that they bear close scrutiny. The species summaries in Chapter 3 describe variations in colouring with age, or variations in bar pattern with size, for example, that may lead to misidentifications.

In the later chapters, we discuss biology of the actinians, biology of the fishes, and how the two influence one another. We conclude with a chapter on keeping these lovely animals in aquaria.

The fish partner may provide an important clue to the identity of the anemone. *Premnas biaculeatus* is only found with *Entacmaea quadricolor*. (G.Allen)

# Sea Anemones

## CLASSIFICATION

Sea anemones are invertebrates, or animals lacking backbones, in contrast with fishes, which are vertebrates. Over 95% of all the kinds of animals in the world are invertebrates, most of them insects.

Primitive animals, anemones belong to the phylum variously known as Cnidaria (with a silent "c") or Coelenterata ("se-len-ter-a'-ta"). The former name alludes to the cnidae, or nematocysts, that are manufactured by all members of this phylum, and only by them. The latter means "hollow gut," referring to the single body cavity that serves as stomach, lung, intestine, circulatory system, and everything else. There is but one opening (the mouth) into this cavity, through which all water, food, and gametes pass in and out. It is surrounded by few or many tentacles, which are finger-like or filamentous projections, typically studded with nematocysts. They are active in capturing food and transferring it to the mouth, and may be used defensively, too.

▲ A diver inspects *Heteractis magnifica* on a coral reef in the Maldive Islands, central Indian Ocean. (H. Debelius/Ikan)

◄ *Heteractis magnifica* with fish symbiont *Amphrprion perideraion*, Great Barrier Reef, Australia. (R. Steene)

Members of the Class Anthozoa (which also includes hard and soft corals), sea anemones live attached to firm objects, generally the sea floor, or embedded in its sediments. An anemone's mouth points generally away from the substratum, and is surrounded by relatively short tentacles. Unlike most other anthozoans, sea anemones lack skeletons of any sort and are solitary. Anthozoans such as corals commonly exist as colonies, with many anemone-like individuals attached to one another. Each cylindrical individual is called a polyp. Members of the other three cnidarian classes may exist as polyps, but additionally (or exclusively) as medusae (singular is medusa). A medusa is little more than an upside-down polyp lacking a skeleton, free to swim in the open sea, with somewhat lengthened tentacles — in short, a jellyfish.

By contrast with their fish symbionts, the 10 host anemones are not all closely related to one another. Belonging to the anthozoan order Actiniaria (hence the term "actinian"), they are members of three different families. The Actiniidae, of which *Entacmaea* and *Macrodactyla* are members, is the largest family of sea anemones, and that to which most common, temperate, shore species belong. The exclusively tropical Stichodactylidae, with the genera *Heteractis* and *Stichodactyla*, is the main host family. Also tropical, the Thalassianthidae contains three genera, including *Cryptodendrum*. Unlike the fishes, in which all members and only members of the damselfish subfamily Amphiprioninae are symbiotic, most members of the families Actiniidae and Thalassianthidae do not participate in symbioses with fishes, and there are also some non-symbiotic stichodactylids.

## *IDENTIFYING SEA ANEMONES*

Nearly all publications on anemone identification are technical. They deal with features such as nature of the animal's muscles, size and distribution of nematocysts, and arrangement of tentacles in relation to internal anatomy. Such characters, which are retained in preserved specimens, require dissection and histological examination to study. They are used partly because most species from the tropics (especially prior to the 20th century) and from deep seas (until the recent advent of submersibles) were originally known only from preserved specimens. We believe actinians can be identified in the field, based on appearance and habitat, although some experts consider nematocyst analysis essential.

A sea anemone is an extremely simple animal. It may be thought of as a cylinder that is closed at both ends. The lower, or pedal, end may be pointed for digging into soft sediments. In anemones of most families, like the host actinians, it is adapted as a pedal disc, which attaches firmly to a solid object like a coral branch or rock (often buried in sediment). In the center of the oral disc, at the opposite, unattached end, is the mouth. Hollow tentacles, arising from the oral disc, surround it. They may be few or many, and arrayed in radial rows or in circlets. Their form is highly diverse — short or long, thin or thick, pointed or blunt, globular or tree-like.

# SEA ANEMONES

## ANATOMY OF A SEA ANEMONE

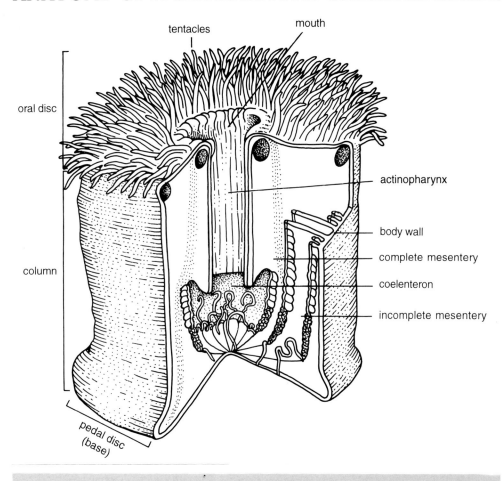

## CROSS SECTION OF AN ANEMONE

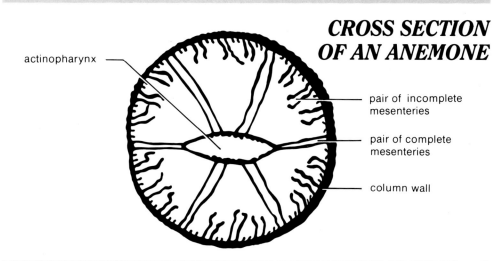

# SEA ANEMONES

Colour and shape are helpful features for identifying sea anemones. *Heteractis magnifica* with fish symbiont *Amphiprion perideraion* on the Great Barrier Reef. (R. Steene)

Tentacle number, form, and arrangement are very important in distinguishing genera and species. The cylindrical column (body) of anthozoans is not completely hollow, the name Coelenterata notwithstanding. In sea anemones, vertical partitions (mesenteries) extend from the column wall across the central space part or all of the way to the throat (actinopharynx). Viewed in cross-section, the column therefore resembles a wheel with spokes. Mesenteries also attach on the underside of the oral disc (the radiating lines of attachment may be visible in an animal that is well expanded, has few tentacles, and/or has a thin oral disc), and tentacles arise between them.

In animals with few tentacles, much of the oral disc, the mouth, and sometimes even the upper end of the throat, into which the mouth opens, are visible. The disc can be radially or circularly patterned; the mouth, which can be circular or elongate, may be elevated on a conical projection and may differ in colour from the oral disc.

The column is appropriately tapered to accommodate a pedal and/or oral disc of smaller or greater diameter than itself. In most species of host actinian, the oral disc is much broader than the column. The column, which may be patterned (commonly splotches of colour or longitudinal stripes), can also bear specialised structures along part or all of its length. For example, some tropical anemones (but none that host clownfishes) have branched projections from the lower column. Most host actinians have, in the upper part, longitudinal rows of small warts (verrucae; singular is verruca) to which particles of gravel may adhere; commonly these are pigmented differently from the rest of the column.

Sea anemone colour pattern can be important for field identification, but colour itself, being highly variable in most actinians, is of little diagnostic value. Symbiotic algae may affect anemone (as well as coral) colour, either by imparting their own golden brown colour, or by stimulating the animal to produce a pigment that protects the algae against excessive sunlight. Therefore, anemones often blend in with corals and with sand, explaining how such large animals may be so difficult to detect in nature.

Presence or absence of verrucae is a character defining genera. Thus, all species in a particular genus do (e.g. *Stichodactyla*) or do not (e.g. *Entacmaea*) have verrucae. Arrangement of tentacles is also important in defining genera. There may be one tentacle per space between mesenteries (so that number of tentacles equals number of mesenteries attaching to the oral disc) or there may be more than one tentacle between each two mesenteries. Members of the family Actiniidae have one tentacle per space. Anemones of the families Stichodactylidae and Thalassianthidae can have so many tentacles because up to several, radially arrayed rows of tentacles arise from alternate spaces (the endocoels), whereas only one tentacle arises from the other spaces (exocoels). The single tentacle is positioned at the very edge of the oral disc (margin). This arrangement may be obvious when the animals are well extended.

Presence or absence of verrucae is important for anemone identification. *Stichodactyla mertensii* has well defined verrucae on the upper part of the column. (R. Eisenhart)

# SEA ANEMONES

Tentacle shape is diagnostic for *Entacmaea quadricolor*, although its tentacles sometimes lack these swellings. (H. Debelius/ Ikan)

## *IDENTIFICATION AIDS*

The following two keys should be used in conjunction with one another. After making an identification by means of the keys, ascertain in the detailed species section that geography and fish are appropriate. Fish symbionts can be a useful character in the field, as mentioned in the last chapter, but are not part of either key.

Primarily features of the oral disc/tentacles are used in the morphological key because little else may be visible (the pedal disc of an anemone living in sediments commonly is attached to a buried object; for one living on a firm substratum it usually is attached within a hole or crevice). The key can even be used (judiciously) to identify animals from photographs. A major caveat is that arrangement and variability of tentacles, in addition to shape, are important, so a photograph of a small area may not provide sufficient information.

In some cases, tentacles alone cannot be conclusive. In all cases, the column should be examined to confirm an identification, if at all possible. This usually requires turning up the edge of the oral disc to reveal, for example, the characteristic longitudinal rows of purple or orange verrucae in *S. mertensii* or the leathery, grey column of *H. crispa*. Such handling may provoke the anemone to retract, partly or completely, into crevice or sand. Sometimes, however, partial retraction reveals precisely the area of interest, and makes for some unusual perspectives.

These keys and descriptions may not work well with captive animals. As explained in Chapter 6, aquarium-kept anemones can lose their colour after a remarkably short time, probably because their algae do not thrive under artificial conditions, and tentacle shape may also change. Fish symbionts should not be used to identify anemones in captivity, as they can be in nature, because many fishes can acclimate to most host anemones.

To use a key, enter at choice 1. There are two alternatives, lettered A and B; select the one that best describes the animal being identified. That is followed by another number. Go to that number, and select from between the alternatives. Ultimately, one of the alternatives will be the name of a species — the identity of the animal in question, if all works as it should. Note that variable species are listed more than once.

## ANATOMICAL KEY

| **1** | **A:** | All tentacles of identical form (although length may vary) .................................................. 3 |
|---|---|---|
| | **B:** | Tentacles of two forms ........................................................ 2 |

| **2** | **A:** | Tentacles long; some finger-like with blunt end, others with bulb at or near end; equator of bulb usually white, tip of tentacle usually red ............................................................. *Entacmaea quadricolor* |
|---|---|---|
| | **B:** | Tentacles extremely short, extraordinarily sticky; in two fields, commonly of different colours: those in band near margin simple elongate bulbs about 1 mm diameter, those covering central oral disc resemble minute stalked gloves .............................. *Cryptodendrum adhaesivum* |

| **3** | **A:** | Most or all tentacles shorter than 30 mm ............................... 9 |
|---|---|---|
| | **B:** | Most or all tentacles longer than 30 mm ................................ 4 |

| **4** | **A:** | Tentacles pointed, relatively few, exposing much of oral disc; tentacle ends may be pigmented; column with verrucae, may be buried in sediment .............. 6 |
|---|---|---|
| | **B:** | Tentacles of uniform diameter (except that tip may be wider), relatively numerous, obscuring much of oral disc; tentacles uniformly pigmented or with pale tips; column with or without verrucae, not buried in sediment ...................................................................... 5 |

| **5** | **A:** | Tentacle tip blunt to slightly swollen, may be different colour than rest of tentacle; occasionally bifurcate. Pedal disc nearly as wide as oral disc; column of uniform bright colour, with obscure verrucae (of same colour; rarely lighter or darker) ..... *Heteractis magnifica* |
|---|---|---|
| | **B:** | Tentacles all finger-like with blunt tips, all with bulb at or near end, or mixture; equator of bulb usually white, tip of tentacle usually red. Pedal disc much narrower than oral disc; column brown, occasionally greenish or reddish, without verrucae *Entacmaea quadricolor* |

| **6** | **A:** | Tentacles smoothly tapered ..................................................... 7 |
|---|---|---|
| | **B:** | Tentacles with swellings at intervals, either entirely around or only on oral surface, so that each may resemble a string of beads; swellings may be white ............................................................. *Heteractis aurora* |

# SEA ANEMONES

| 7 | A: | Oral disc with bold, light coloured, radial stripes that may extend up at least inner tentacles; long tentacles sparse, may assume corkscrew form; upper column dark, with light-coloured round to eye-shaped varrucae ................................. *Macrodactyla doreensis* |
|---|---|---|
|   | B: | Oral disc without light coloured stripes that extend up some of tentacles; tentacles long to relatively short .................................................................................... 8 |

| 8 | A: | Column white to grey (at least in upper part), texture thick and leathery, widely flared at oral end, with many verrucae per row; tentacles normally long, abundant. ............................... *Heteractis crispa* |
|---|---|---|
|   | B: | Column violet brown on upper part, relatively thin, more or less uniformly tapered, with relatively few verrucae per row; tentacles short, sparse ........ *Heteractis malu* |

| 9 | A: | Tentacles all approximately same length and very adhesive (may pull off the anemone on contact with a person's fingers); oral disc diameter commonly to .5-.8 m; non-adhesive verrucae on upper column blue, purple, maroon, red, or same colour as column ................................................................................... 10 |
|---|---|---|
|   | B: | Patches of tentacles, in center or elsewhere, may be much longer than others, not strongly adhesive; oral disc to 1 m or more diameter, flaccid, held in flat, expanded position by adhesion of prominent orange or magenta verrucae to underlying substratum; narrow column grey to white ............. *Stichodactyla mertensii* |

| 10 | A: | Tentacles tapered to blunt point, all constantly vibrating; tentacles may be brilliant blue or deep purple; exocoelic tentacles not noticeably more robust than endocoelic. Column relatively narrow ................................................... *Stichodactyla gigantea* |
|----|----|---|
|    | B: | Each tentacle with narrow stalk and pointed to globose end, not constantly vibrating; tentacles may vary in colour so appearance of oral disc is variegated; exocoelic tentacles more robust than endocoelic. Column substantial ............................ *Stichodactyla haddoni* |

## HABITAT KEY

| 1 | A: | Living on/in sediment, e.g. sand, mud ...................... 8 |
|---|---|---|
|   | B: | Living on a firm substratum, e.g. rocks, coral ........... 2 |

# SEA ANEMONES

| | | |
|---|---|---|
| **2** | **A:** | Pedal disc attached in crevice or hole, so it and most or all of column hidden ................................................... 5 |
| | **B:** | Pedal disc attached at surface of substratum, so it and most or all of column normally visible .................... 3 |
| **3** | **A:** | Usually in prominent position, such as elevated plate coral, or top of boulder, occasionally on branching coral; column about as broad as pedal disc, of uniform bright colour; tentacle tips blunt to slightly swollen .................................................................. *Heteractis magnifica* |
| | **B:** | Pedal disc wrapped around branch of coral such as *Acropora* ................................................................. 4 |
| **4** | **A:** | Solitary. Column white to grey with prominent verrucae, leathery texture, widely flared at oral end; tentacles long, abundant, tapered to points, their tips often mauve or blue ....................... *Heteractis crispa* |
| | **B:** | At least several individuals beside one another; columns brown, occasionally greenish or reddish, without verrucae. Each brown tentacle finger-like with blunt end or with bulb at or near end or both; equator of bulb usually white, tip of tentacle usually red .......................................... *Entacmaea quadricolor* |
| **5** | **A:** | Solitary; column commonly extends horizontally from hole ................................................................................ 6 |
| | **B:** | At least several individuals beside one another; extend vertically from crevice in top of reef. Only long tentacles protrude above surface of reef; finger-like with blunt end or with bulb at or near end or both; equator of bulb usually white, tip of tentacle usually red .......................... *Entacmaea quadricolor* |
| **6** | **A:** | Oral disc much broader than column, spreads flat over surface of substratum, or may be partly propped against vertical surfaces; covered with mostly very short tentacles ................................................................ 7 |
| | **B:** | Oral disc slightly broader than column; relatively few, long tentacles, each usually with bulb near end, red tip. Commonly on reef slope ......... *Entacmaea quadricolor* |
| **7** | **A:** | Flaccid oral disc held in expanded position by adhesion of prominent orange or magenta verrucae to underlying substratum. Tentacles all finger-like; most short but some, in center of oral disc or elsewhere, may be much longer ......... *Stichodactyla mertensii* |

# SEA ANEMONES

| | | |
|---|---|---|
| | **B:** | Tentacles extremely short, extraordinarily sticky; in two fields, commonly of different colours: those in band near margin simple elongate bulbs about 1 mm diameter, those covering central oral disc resemble stalked gloves ............... *Cryptodendrum adhaesivum* |
| **8** | **A:** | Away from immediate vicinity of reefs; in mud or clean sand .......................................................... 12 |
| | **B:** | In patches of sand among reefs, or in sand mixed with rubble. ................................................... 9 |
| **9** | **A:** | Oral disc visible among relatively few, long tentacles ................................................................ 10 |
| | **B:** | Oral disc deeply folded, densely covered with short tentacles, each tapered to a blunt point, all constantly vibrating ............................. *Stichodactyla gigantea* |
| **10** | **A:** | Tentacles smoothly tapered ..................................... 11 |
| | **B:** | Tentacles with swellings at intervals, either entirely around or only on oral surface, so that each may resemble a string of beads; swellings may be white ....................................................... *Heteractis aurora* |
| **11** | **A:** | Column thin, with few verrucae per row; tentacles short and sparse ............................. *Heteractis malu* |
| | **B:** | Column thick and leathery, white to grey, with prominenet verrucae; tentacles abundant, long .................................................................. *Heteractis crispa* |
| **12** | **A:** | Oral disc usually undulating, at surface of clean sand, capable of withdrawing completely into sand; each tentacle with narrow stalk and pointed to globose end, may vary in colour so appearance of oral disc is variegated; exocoelic tentacles conspicuous, more robust than endocoelic; column pale with either pale, non-contrasting or light rose to purplish verrucae ............................. *Stichodactyla haddoni* |
| | **B:** | Oral disc flat, at surface of mud or muddy sand; with radial stripes that may extend up at least inner tentacles; tentacles long, evenly tapered to point, sometimes assuming corkscrew shape; upper column dark, with light-coloured round to eye-shaped verrucae .............................. *Macrodactyla doreensis* |

*Stichodactyla mertensii* with pair of *Amphiprion chrysopterus*, Madang, Papua New Guinea. (G. Allen)

# *Cryptodendrum adhaesivum*
## KLUNZINGER, 1877
## ADHESIVE SEA ANEMONE

*Original description:*

As *Cryptodendrum adhaesivum*, from a specimen collected at Koseir in the Red Sea.

*Other name previously used:*

*Stoichactis digitata* (by Doumenc 1973).

*Diagnostic field characters:*

Tentacles extremely sticky; short (to 5 mm long), dense, of two forms; those in centre of oral disc have narrow stalk with five or more short branches at end (i.e. resembling a miniature glove); those near the edge simple elongate bulbs about 1 mm diameter; at extreme margin is a ring of tentacles like the central ones but with fewer branches. Tentacles of the two forms usually different colours: observed combinations include yellow and pink, blue and grey, green and brown; occasionally tentacles of another colour occur in patches amid those of predominant colour.

*Details:*

Oral disc to 300 mm diameter, flat when expanded, but commonly undulating. Entirely covered with tentacles except immediately around mouth, which can be fuchsia, yellow, green, white. Moreover, tentacle stalk and tips may differ in colour. Therefore, may be extremely colourful animal.

*Similar species:*

Specimens of *Stichodactyla* are superficially similar, with many, short tentacles. However, the two distinct types of tentacles arrayed in separate fields is a feature unique to *C. adhaesivum*. Tentacles of other species may adhere, and pull off the anemone; those of *C. adhaesivum* remain attached to the anemone.

*Distribution:*

Australia to southern Japan and Polynesia, Micronesia, and Melanesia westward to Thailand, Maldives, and the Red Sea.

*Fish symbiont*

A. clarkii

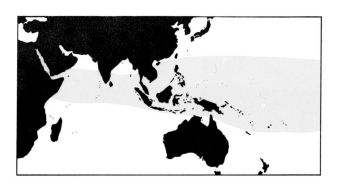

# SEA ANEMONES

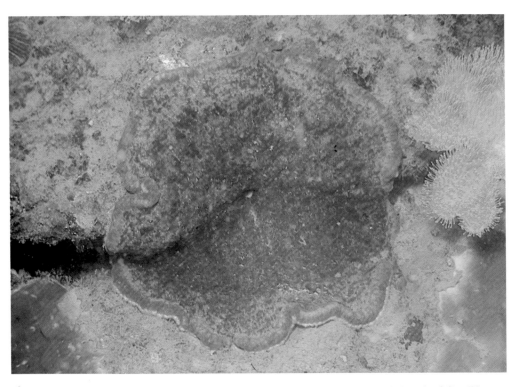

▲ *C. adhaesivum*, Madang, Papua New Guinea. (G. Allen)

▽ *C. adhaesivum*, close up of edge of oral disc. (H. Debelius/Ikan)

# SEA ANEMONES

## *Entacmaea quadricolor*
(RÜPPELL AND LEUCKART, 1828)
### BULB-TENTACLE SEA ANEMONE

*Original description:*

As *Actinia quadricolor*, from specimens collected in the Red Sea, near Suez.

*Other names previously used:*

*Gyrostoma heliant[h]us* (by Fishelson 1970, Masry 1971, Fricke 1974), *G. quadricolor* (by Fishelson 1970, Fricke 1974), *Physobrachia ramsayi* (by Mariscal 1970, Friese 1972, Mariscal 1972, Uchida *et al.* 1975), *P. douglasi* (by Allen 1972, 1975, 1978, Ross 1978, Cutress and Arneson 1987), *Radianthus gelam* (by Allen 1972, 1978, Friese 1972), *Cymbactis actinostoloides* (by Moyer and Sawyers 1973), *Parasicyonis actinostoloides* (by Uchida *et al.* 1975; as *P. actinostoroides*, Moyer 1976, Moyer and Bell 1976), *P. maxima* (by Uchida *et al.* 1975, Moyer 1976, Moyer and Bell 1976).

*Diagnostic field characters:*

Each long (to 100 mm) brown tentacle usually with bulb at or somewhat below end; tip of tentacle red (rarely blue), equator of bulb white. Bulb seems to be related to presence of fish, and can disappear; tentacle lacking a bulb has white ring where equator would form. Tentacles without bulbs are blunt-ended. As a rule, in shallow water (e.g. on tops of reefs), polyps small (oral disc diameter 50 mm), clustered together in crevices or adjacent on coral branches, so that tentacles are confluent, forming extensive field; in deep water (e.g. on reef slopes), polyps solitary, large (to 400 mm diameter), with base anchored in deep hole.

*Details:*

Animal commonly attached deeply in crevice or hole so that only emergent tentacles visible. Column without verrucae; usually brown, sometimes reddish or greenish; gradually flared from small pedal disc. Oral disc same brown colour as tentacles. Tentacles can collapse when disturbed, appear grey-green. The most numerous host actinian, widespread geographically and abundant locally.

*Similar species:*

The smooth column is unique among symbiotic sea anemones, as are the bulbed tentacles.

*Distribution:*

Micronesia and Melanesia to East Africa and the Red Sea, and from Australia to Japan.

*Fish symbionts:*

A. akindynos

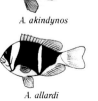
A. allardi

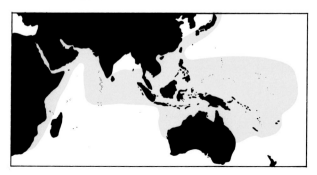

# SEA ANEMONES

▲ Clonal form of *E. quadricolor*, Coral Sea, Australia. Symbiotic fish is *A. melanopus*. (H. Horn)

Solitary form of *E. quadricolor*, close-up of tentacles, Madang, Papua New Guinea. (R. Steene) ▼

*Fish symbionts:*

*A. bicintus*

*A. chrysopterus*

*A. clarkii*

*A. ephippium*

*A. frenatus*

*A. mccullochi*

*A. melanopus*
(primarily clustered form)

*A. omanensis*

*A. rubrocinctus*

*A. tricinctus*

*Premnas* (solitary form only)

# *Heteractis aurora*
(QUOY AND GAIMARD, 1833)
## BEADED SEA ANEMONE

*Original description:*
As *Actinia aurora*, from specimens collected in New Ireland.

*Other names previously used:*
*Radianthus koseirensis* (by Mariscal 1970, 1972), *Radianthus simplex* (by Allen 1972, Moyer 1976), *Bartholomea* sp. (by Uchida 1975).

*Diagnostic field characters:*
Tentacles to 50 mm long, with swellings (up to 20 on long tentacles) at intervals, either on only oral side or nearly surrounding a tentacle so that it resembles a string of beads. Swellings often white. Oral disc broad, to 250 mm or possibly more, spread flat or slightly undulating at surface of sediment.

*Details:*
Tentacles brown or purplish, arising from oral disc of the same colour; outermost tentacles may be shorter than inner, and can have purplish or greenish cast. Oral disc mostly visible because of sparse tentacles; may have white or brown radial markings that can continue onto tentacles. Tentacles may be sticky to touch; tapered to point that may be magenta in colour. Adhesive verrucae on upper column lighter in colour than column; lower column often mottled or solid orange or red. Animals attached to buried objects capable of retracting completely into sediment.

*Similar species:*
*Macrodactyla doreensis*, *Heteractis malu*, and *H. crispa* also live burrowed into sediment. Tentacles of the other two species of *Heteractis* may also be magenta-tipped, but those of *H. aurora* are unique in having swellings at intervals. Tentacles of *H. aurora* are intermediate in length between those of *H. malu* (shorter) and *H. crispa* and *M. doreensis* (longer). Tentacles in some individuals of *H. aurora* are nearly as sparse as are those of *H. malu*. The column of *H. aurora* is similar in texture to that of *H. malu*.

*Distribution:*
Micronesia and Melanesia to East Africa and the Red Sea, and Australia to the Ryukyu Islands.

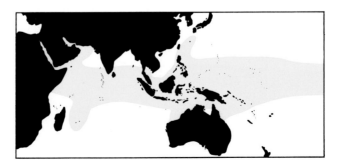

# SEA ANEMONES  I

▲ *H. aurora* (with juvenile *A. clarkii*), Flores, Indonesia. (G. Allen)

Tentacle shape of *H. aurora* is very distinctive. (R. Steene) ▼

*Fish symbionts:*

*A. akindynos*

*A. allardi*

*A. bicinctus*

*A. chrysogaster*

*A. chrysopterus*

*A. clarkii*

*A. tricinctus*

# SEA ANEMONES

## *Heteractis crispa*
(EHRENBERG, 1834)
### LEATHERY SEA ANEMONE

*Original description:*

As *Actinia crispa*, from specimens collected in the Red Sea.

*Other names previously used:*

*Radianthus kuekenthali* (by Mariscal 1970, 1972, Uchida *et al.* 1975, Moyer 1976), *R. malu* (by Allen 1972, 1973, 1975), *R. ritteri* (by Allen 1978), *H. macrodactylum* (by Cutress and Arneson 1987).

*Diagnostic field characters:*

Tentacles long (typically to 100 mm), sinuous, evenly tapered to point, often tipped mauve or blue, rarely yellow or green. Oral disc widely flared, may exceed 500 mm diameter, but commonly 200 mm. Column grey in colour, leathery in texture, with prominent adhesive verrucae; lower part rarely mottled with yellow. Column buried in sediment so oral disc lies at surface of sediment, or pedal disc attached to branching coral.

*Details:*

Tentacles very numerous — to 800 counted. Oral disc usually brownish violet, or grey, rarely bright green. Tentacles shrivel when animal is disturbed, and assume green or grey luster; may shorten greatly in absence of fish. For an animal attached to coral branches, verrucae adhere to branches, holding oral disc open among them; verrucae adhere to sediment particles if animal lives in sediment.

*Similar species:*

*Heteractis magnifica* rarely lives on branching coral. Its blunt tentacles and brightly coloured column are distinctive. *Macrodactyla doreensis*, *Heteractis aurora*, and *H. crispa* also live burrowed into sediment. In contrast to anemones of those three species, *H. crispa* has many, long tentacles. Those of the other two burrowing species of *Heteractis* may also be magenta-tipped. Tentacles of *H. crispa* may contract in the absence of a fish, but they are more numerous than in *H. malu*, and lack the swellings of *H. aurora*. The column of *H. crispa* is unique among host actinians.

*Distribution:*

French Polynesia, Micronesia, and Melanesia to the Red Sea, and Australia to Japan.

*Fish symbionts:*

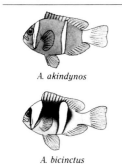

*A. chrysopterus*

*A. clarkii*

# SEA ANEMONES

*Fish symbionts:*

*A. ephippium*

*A. latezonatus*

*A. leucokranos*

*A. melanopus*

*A. omanensis*

*A. percula*

*A. perideraion*

*A. polymnus*

*A. sandaracinos*

*A. tricinctus*

▲ *H. crispa*, Madang, Papua New Guinea. (G. Allen)

This unusual and beautiful form of *H. crispa* is sometimes encountered in the Maldive Islands. The fish is *A. clarkii*. (J. Neuschwander/Ikan) ▼

# *Heteractis magnifica*
(QUOY AND GAIMARD, 1833)
## MAGNIFICENT SEA ANEMONE

*Original description:*

*Actinia magnifica*, from specimens collected at Vanikoro, Santa Cruz Islands, New Hebrides.

*Other names previously used:*

*Radianthus ritteri* (by Mariscal 1970, 1972, Allen and Mariscal 1971, Allen 1972, 1975, 1978), *R. paumotensis* (by Allen 1972, Friese 1972), *R. macrodactylus* (by Uchida *et al.* 1975), *R. malu* (by Allen 1978), *H. ritteri* (by Cutress and Arneson 1987).

*Diagnostic field characters:*

Typically occupies fully exposed, prominent position, attached to solid object such as coral boulder. Cylindrical column of uniform bright colour (commonly blue, green, red, white, chestnut brown). Oral disc to 1 m diameter (although commonly 300-500 mm), flat to gently undulating, densely covered with finger-like tentacles (to 75 mm long) that hardly taper to blunt or slightly swollen end. May irritate human skin and raise welts.

*Details:*

Lower portion of tentacles same colour as oral disc (usually shade of brown), terminal portion yellow, green, or white; some tentacles bifurcate or with side branch. Tentacles approach mouth to within 20-30 mm; central oral disc yellow, brown, or green, often raised so that mouth sits on a cone. Column with longitudinal rows of translucent verrucae same colour as column or slightly lighter or darker. Animal capable of almost complete contraction so that only a tuft of tentacles is visible in center.

In western Indonesia, several small individuals of identical colouration may cluster together, resembling one large animal. Elsewhere (e.g. Maldives, Malaysia, French Polynesia), tens or hundreds of identically coloured individuals form extensive beds; presumably they constitute a clone.

*Similar species:*

This is probably the most distinctive and most commonly photographed species of host actinian. Its exposed habitat is unique, as is its brightly coloured, gently flared column. Only *Stichodactyla mertensii* may exceed it in diameter, but *H. magnifica* is a much more substantial animal. Its blunt tentacles are unique in the genus; those of *S. haddoni* are similarly shaped but shorter and more densely arrayed.

# SEA ANEMONES

*Distribution:* French Polynesia to East Africa, and Australia to the Ryukyu Islands.

*Fish symbionts:*

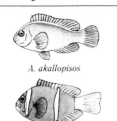

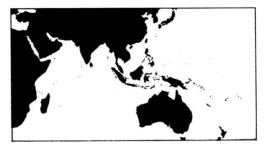

A. akallopisos

A. akindynos

A. bicinctus

A. chrysogaster

A. chrysopterus

A. clarkii

A. leucokranos

A. melanopus

A. nigripes

A. ocellaris

A. percula

A. perideraion

▲ *H. magnifica* (with *A. perideraion*), Great Barrier Reef. (L. Preston)

▲ *H. magnifica* in a contracted state, Great Barrier Reef. The fish is *A. perideraion*. (R. Steene)

# *Heteractis malu*
## (HADDON AND SHACKLETON, 1893)
### DELICATE SEA ANEMONE

*Original description:*

As *Discosoma malu*, from specimens collected at Mer, in the Torres Straits.

*Other names previously used:*

*Macranthea cookei* (by Reed 1971), *Radianthus papillosa* (by Dunn 1974, Moyer 1976), *Antheopsis papillosa* (by Cutress 1977).

*Diagnostic field characters:*

Tentacles sparse, stubby (rarely to 40 mm long), of variable length even within one radial row, commonly magenta-tipped. Oral disc lies at surface of sediment in which delicate column is burrowed. Column commonly pale cream or yellow colour, may have splotches of deep yellow or orange.

*Details:*

Tentacles arise from brown or purplish (rarely bright green) oral disc as much as 200 mm in diameter that may have white radial markings; evenly tapered to point or slightly inflated in middle; lower part same colour as oral disc, but upper portion may have several white rings or green end. Column very thin in expansion; upper part violet-brown (due to zooxanthellae) with longitudinal rows of adhesive verrucae. Anemones can retract completely into sediment; most common in shallow, quiet waters.

*Similar species:*

*Macrodactyla doreensis*, *Heteractis aurora*, and *H. crispa* also live burrowed into sediment. The columns of all four species may have red or yellow blotches; that of *H. crispa* is much firmer than that of *H. malu* (which is similar to that of *H. aurora*). In contrast to animals of the other three species, *H. malu* has relatively short tentacles, typically of variable length; tentacles of the other two species of *Heteractis* may also be magenta-tipped. Tentacles of *H. crispa* may contract in the absence of a fish, but *H. malu* has fewer tentacles per radial row; tentacles of *H. aurora* may be similarly sparse, but have swellings at intervals.

*Distribution:*

Scattered localities from the Hawaiian Islands to Australia and northwards to Japan.

*Fish symbiont:*

A. clarkii

# SEA ANEMONES

▲ *H. malu*, Hawaiian Islands. (A. Reed)

▽ *H. malu* with juvenile *A. clarkii*, Madang, Papua New Guinea. (G. Allen)

# Macrodactyla doreensis
## (QUOY AND GAIMARD, 1833)
## CORKSCREW TENTACLE SEA ANEMONE

*Original description:*

As *Actinia doreensis*, from specimens collected at Port Dorey, New Guinea, now Manokwari, Irian Jaya, Indonesia.

*Other names previously used:*

*Macrodactyla gelam* (by Mariscal 1972), *Radianthus malu* (by Moyer 1976), *H. gelam* (by Cutress and Arneson 1987).

*Diagnostic field characters:*

Tentacles few, long (to 175 mm), all alike, sinuous, evenly tapered to point, sometimes (but not invariably) assuming corkscrew shape. Oral disc widely flared, to 500 mm diameter but commonly considerably less, with radial white lines that may extend onto tentacles; lies at surface of sediment. Column buried in sediment; lower part dull orange to brilliant red, upper part brownish with non-adhesive, prominent white round to ovoid (eye-shaped) verrucae in longitudinal rows.

*Details:*

Oral disc usually purplish-grey to brown, sometimes with a green cast; tentacles basically same colour but tips may be darker or lighter. When disturbed, tentacles may shrivel or may adhere to collector's hand and pull off. Often found in mud, generally no deeper than 5 m, commonly without fish; can retract completely into sediment.

*Similar species:*

*Heteractis aurora*, *H. crispa*, and *H. malu* also live burrowed into sediment, and may have red or yellow pigmentation on the lower column. Tentacles of *M. doreensis* lack bulges, in contrast to those of *H. aurora*, are fewer than those of *H. crispa*, and are longer than those of *H. malu*. The column of *M. doreensis* is thin, and has distinctive verrucae. The distribution of this species is the most restricted of any host anemone.

*Distribution:*

Japan south to New Guinea and northern Australia.

*Fish symbionts:*

A. chrysogaster

A. clarkii

A. perideraion

A. polymnus

# SEA ANEMONES

▲ *M. doreensis*, closeup of tentacles. (G. Allen)

▼ *M. doreensis* with *A. clarkii* and *D. trimaculatus*, Flores, Indonesia. (G. Allen)

# SEA ANEMONES

## *Stichodactyla gigantea*
(FORSSKÅL, 1775)
### GIGANTIC SEA ANEMONE

**Original description:**
As *Priapus giganteus*, from specimens collected in the Red Sea.

**Other names previously used:**
*Discosoma giganteum* (by Gohar 1948, Schlichter 1968), *Stoichactis kenti* (by Mariscal 1969, 1970, 1972; Allen 1972, 1973, 1978; Uchida et al. 1975).

**Diagnostic field characters:**
Deeply-folded oral disc (more pronounced with size), covered with short (average 10 mm), slightly tapering tentacles that typically all vibrate constantly. Tentacles extremely sticky in life, adhering to collector's hand, and pulling off in clumps; but do not cause stinging sensation. Typically in such shallow water that animals may be exposed at low tide.

**Details:**
May be extraordinarily abundant. Oral disc rarely as much as 500 mm diameter, usually lies at surface of sand, often among corals; pedal disc attached to buried object. Non-adhesive verrucae on upper column blue to maroon, contrasting with yellowish, pinkish, tan, greenish-blue, or grey-green column. Basal portion of each tentacle colour of the oral disc (often tan or pink); colour of bluntly pointed distal part — which is what is generally noted as tentacle colour — commonly brown or greenish, rarely a striking purple, pink, deep blue, or bright green. Much of central oral disc bare, but deep oral disc folds may hide mouth.

**Similar species:**
The shallow, sandy habitat is unusual. *Stichodactyla mertensii*, which lives on hard substrata, has a flat oral disc and distinctive column. *Stichodactyla haddoni* typically lives in cleaner sand and deeper water, and its oral disc folds are more regular and more separate than those of *S. gigantea*; tentacle form and fish symbionts of the two also clearly separate them.

**Distribution:**
Micronesia to the Red Sea, and Australia to the Ryukyu Islands.

# SEA ANEMONES

*Fish symbionts:*

*A. akindynos*

*A. bicinctus*

*A. clarkii*

*A. ocellaris*

*A. percula*

*A. perideraion*

*A. rubrocinctus*

▲ *S. gigantea*, Madang, Papua New Guinea. (G. Allen)

Colourful variant of *S. gigantea*, Madang, Papua New Guinea. Fish is *A. percula*. ▼
(G. Allen)

# *Stichodactyla haddoni*
(SAVILLE-KENT, 1893)
## HADDON'S SEA ANEMONE

*Original description:*

As *Discosoma haddoni*, from specimens collected on the northern Great Barrier Reef.

*Other names previously used:*

*Stoichactis kenti* (by Friese 1972, Moyer and Sawyers 1973), *S. gigantea* (by Friese 1972), *S. haddoni* (by Uchida et al. 1975, Moyer 1976, Moyer and Steene 1979).

*Diagnostic field characters:*

Slightly to deeply folded oral disc lies on or above sand surface; tentacles either bulbous or with basal "stalk," at the end of which is a blunt or swollen terminal portion that can appear puckered (on close examination). Exocoelic tentacles more robust than the endocoelic ones with which they alternate. Column sturdy.

*Details:*

Oral disc diameter commonly 500 mm, rarely 800 mm; yellowish to orange tentacle-free oral area 10-20 mm in diameter. Oral disc, lower portion of tentacles, and column drab — commonly yellowish or tan. Tentacle ends can be green, yellow, grey, or rarely pink, which can give oral disc a variegated appearance. Exocoelic tentacles usually white, may be up to twice as long as endocoelic, point outward in well expanded animals. Tentacles sticky to touch, may adhere to human skin so strongly that they pull off the anemone; contact with them painless but can raise welts. Small, non-adhesive verrucae on uppermost column are same colour as column or light rose to purple. Anemone can pull rapidly and completely beneath the sand when disturbed, leaving its fish to hover over resulting depression.

*Similar species:*

*Stichodactyla gigantea* also lives in sand but typically in shallower water, and folds of its oral disc are less regular and more closely spaced. The oral disc of *S. mertensii*, which lives on firm substrata, lies fairly flat. The column of *S. haddoni* is more substantial than that of either, and its tentacles are longer and distinctively shaped. The other species lack robust exocoelic tentacles. Tentacles pull off of *S. gigantea* as well, but in massive clumps rather than one or several at a time.

*Distribution:*

Fiji Islands to Mauritius, and Australia to the Ryukyu Islands.

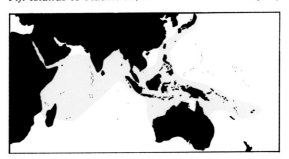

# SEA ANEMONES I

*Fish symbionts:*

A. akindynos

A. chrysogaster

A. chrysopterus

A. clarkii

A. polymnus

A. sebae

A. omanensis

▲ *S. haddoni* with *A. polymnus*, Madang, Papua New Guinea. (G. Allen)

*S. haddoni* with variegated pattern, Solomon Islands. The fish is *A. clarkii*. (R. Steene) ▼

# SEA ANEMONES

# *Stichodactyla mertensii*
## BRANDT, 1835
## MERTENS' SEA ANEMONE

**Original description:**

As *Stichodactyla mertensii*, from specimens collected in the easternmost Caroline Islands.

**Names previously used:**

*Stoichactis giganteum* (by Mariscal 1970, Allen and Mariscal 1971, Allen 1972, 1973, 1975).

**Diagnostic field characters:**

Oral disc to 1 m or even more diameter; tan to white column with longitudinal rows of verrucae pigmented magenta or orange (which appear purplish at depth); non-adhesive tentacles club-shaped to finger-like — all may be short (10-20 mm long), or some (in patches) very long (to 50 mm or more).

**Details:**

This anemone holds the record for oral disc diameter (shape is often more ovoid than circular). Broadly flared oral disc lies smoothly over substratum, following its contours, or undulating slightly, held open by verrucae adhering to underlying coral or rocks, which may be related to this anemone's living only on hard surfaces, often reef slopes. No verrucae below wide upper column, but splotches of pigment continue down short, narrow column in more or less longitudinal streaks. Small pedal disc frequently attached in crevice into which animal can retract (although not rapidly). Tentacles, of uniform diameter, blunt-ended or pointed: short ones same colour as the oral disc, sometimes with narrow white basal portion; long ones may be white-, yellow-, or green-ended. Tan oral disc almost entirely covered with tentacles; yellow or greenish tentacle-free oral area 20-50 mm diameter. Symbiotic fishes may be melanistic.

**Similar species:**

*Stichodactyla gigantea* and *S. haddoni* live in sand. Their oral discs are wavy, in contrast with that of *S. mertensii*, and their columns more substantial. The only other host anemone that rivals *S. mertensii* in size is *H. magnifica*, which has longer, blunt tentacles and a brightly coloured, cylindrical column.

**Distribution:**

Micronesia and Melanesia to East Africa, and Australia to the Ryukyu Islands.

**Fish symbionts:**

*A. akallopisos*

*A. akindynos*

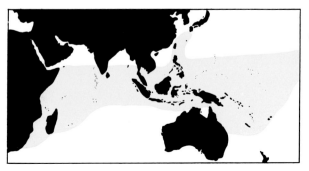

*A. bicinctus*

# SEA ANEMONES

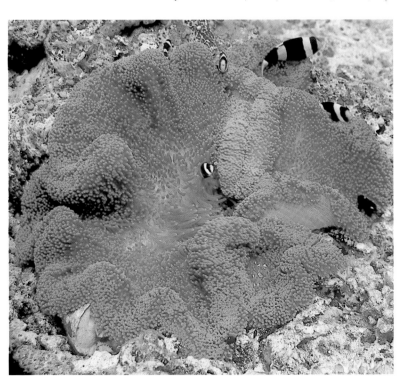

S. *mertensii* with A. *akindynos*. (H. Debelius/Ikan)

Variety of S. *mertensii*, Flores, Indonesia. (G. Allen)

*Fish symbionts:*

A. *allardi*

A. *chrysogaster*

A. *chrysopterus*

A. *clarkii*

A. *fuscocaudatus*

A. *latifasciatus*

A. *leucokranos*

A. *ocellaris*

A. *sandaracinos*

A. *tricinctus*

# Anemonefishes

## CLASSIFICATION

Anemonefishes are members of the Pomacentridae, one of the largest families in the order Perciformes, with approximately 325 species. Members of this family, commonly known as damselfishes, are almost entirely marine (three mainly brackish water species sometimes occur in fresh water), and most species occur in tropical and, to a lesser degree, subtropical latitudes. About 70% of damselfishes, including anemonefishes, are restricted to the vast Indo-West Pacific region.

*Amphiprion*, and *Premnas* constitute the Amphiprioninae, one of four pomacentrid subfamilies. The only other damselfishes that sometimes dwell with anemones are *Dascyllus trimaculatus* and *D. albisella*, which belong to another subfamily, the Chrominae.

▲ *Amphiprion percula* with *Heteractis magnifica*, Solomon Islands. These fish have an aberrant colour pattern, lacking the normal complete bar across the tail base. (E. Robinson)

◀ A pair of *Amphiprion nigripes*, tending their nest, Maldive Islands. (G. Allen)

## IDENTIFICATION

Colour pattern is the most important feature for identifying an anemonefish in the field. Other useful characters, which include tooth shape, scalation of the head, and body proportions, are mainly of value to laboratory workers identifying preserved specimens. Therefore, the key below to the 27 species of *Amphiprion* and the single species of *Premnas* is based almost entirely on adult colour characteristics. It includes common geographic colour variants that occur in some of the wider ranging species such as *A. clarkii* and *A. melanopus*.

Correct identification of juveniles is often difficult, due to great similarities among many species and to the colour patterns of many differing from those of adults. Therefore, juveniles are not included in the key. The easiest juveniles to identify are those that resemble the adult and that mostly have either no bars or a single white bar: *A. akallopisos, A. leucokranos, A. nigripes, A. perideraion,* and *A. sandaracinos*. Distinctive shape and colouration make the young of *A. latezonatus, A. ocellaris-A. percula* (which, however, are difficult to distinguish from one other), *A.*

Juvenile *Amphiprion nigripes* with *Heteractis magnifica*, Sri Lanka. (G. Allen)

# ANEMONEFISHES

Juvenile to adult colour pattern changes for selected species of *Amphiprion*: from top to bottom — *A. bicinctus*; *A. chrysogaster*; *A. ephippium*; and *A. melanopus*.

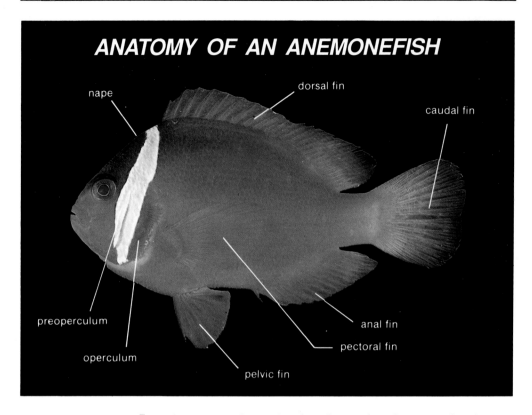

*polymnus*, *A. sebae*, and *P. biaculeatus* relatively easy to identify. More difficult to separate are the red-finned species, whose adults possess a single bar (*A. frenatus*, *A. melanopus*, *A. rubrocinctus*, and *A. ephippium*). Their juveniles generally have two or three white bars, which they lose sequentially from the tail end as they grow. Perhaps most difficult to distinguish are juveniles of species that possess two or three bars as adults and have variable amounts of yellow or orange on body and fins. Curiously, juveniles of these fishes frequently have one bar to begin with, developing the other(s) as they grow. For many such species, it may be necessary to observe a transitional series of progressively larger fish in order to link the smallest young with adults. Fortunately, in field situations this is often possible, as "family" groups of a single species composed of individuals of increasing size usually occur with larger anemones. The accompanying illustration showing the change in colour pattern between juveniles and adults is intended to serve as a general guide in assisting with the identification of juvenile and intermediate stages.

The key is followed by individual species accounts that provide additional information on important colour pattern features, including an underwater photograph, means of distinguishing similar species, anemone host(s), and details of distribution. Several "species pairs" share nearly identical patterns: for example, *Amphiprion allardi*-*A. chrysopterus*, *A. chrysogaster*-*A. tricinctus*, and *A. akindynos*-*A. chagosensis*. Fortunately, the members of each pair have widely separated geographic ranges. Thus, knowledge of the distribution of a species may be very useful in identifying it correctly. *Dascyllus trimaculatus*, juveniles of which are often encountered with anemones, is also included among the species accounts, but is omitted from the key as it is easily distinguished from *Amphiprion* and *Premnas*.

External anatomy of an anemonefish (*Amphiprion melanopus*, Fiji Islands). (J. Randall)

## *COLOUR VARIATION*

The colour of anemonefish of one species sometimes varies. Geographic variation is most common among widely distributed species. For example, *A. clarkii*, which has the broadest distribution of any anemonefish, is exceedingly variable over its range. Some of the commonly encountered colour varieties are shown in the accompanying illustration. Another type of variation is melanism, (black pigmentation), which is somehow induced by the host anemone. This topic is discussed in more detail in Chapter 5, where several examples are illustrated.

Other categories of variation are related to sex (see Chapter 4), are due to hybridisation, or are purely random. Hybrid crosses between closely related species have been produced in aquarium conditions, and at least one probable cross, involving *A. chrysopterus* and *A. leucokranos*, has been observed by us in Papua New Guinea. The most common sort of random variation involves irregularities in the shape of the white bars displayed by most species, particularly the head bar. In some cases, one or more bars may be absent or greatly abbreviated. This type of variation is evident in the unusually marked specimens of *A. percula* that are illustrated at the beginning of this chapter and in Chapter 6.

Probable hybrid between *Amphiprion chrysopterus* and *A. leucokranos*, Madang, Papua New Guinea. The anemone is *Heteractis crispa*. (G. Allen)

# ANEMONEFISHES

Some of the many colour varieties of *Amphiprion clarkii*: left to right and top to bottom — Vanuatu, Maldive Islands, New Guinea, Okinawa, Andaman Sea, Christmas Island (Indian Ocean), Western Australia and Java Sea.

# ANEMONEFISHES

▲ Colour pattern difference is related to sex in some anemonefishes. The female of each of the following pairs appears on top: left to right — *Premnas biaculeatus, Amphiprion clarkii,* and *A. frenatus.*

# KEY TO ANEMONEFISHES

## *HOW TO USE THE KEY —*

The key is divided into groups based on the number of (vertical) white bars possessed by subadult and adult anemonefish: (A) no bars; (B) one bar; (C) two bars; and (D) three bars. The first step is to determine where to enter the key, based on pattern of the fish being identified. Subsequently, each section contains sets of two choices (couplets) alternately labeled 1 or 2 and A or B. Eliminate the choices that do not agree with the fish in question and follow the alternative path until the correct identification is made. The accompanying black and white drawings illustrate common variants, greatly assisting in the identification process. As mentioned previously, this key relies largely on colour pattern, and avoids technical terms.

As a practice exercise, assume that you are attempting to identify the common species that occurs in the Maldive Islands, *A. nigripes*. The fish is characterised by a single, narrow bar just behind the eye, reddish brown colour, and black on the breast and belly extending onto the pelvic and anal fins. As there is only a single white bar, the correct section of the key is Group B. The first two choices in this section are (1) pelvic and/or anal fins dark, and (2) pelvic and anal fins light. The fish in question agrees with (1); therefore, read choices, (A) and (B) in this section. The first choice under (A), "head bar narrow, equal to or less than width of eye; black colour on body restricted to breast and belly" is descriptive for the fish in question and, indeed, the drawing of *A. nigripes* that illustrates this entry in the key agrees well with the fish. Conversely, the descriptive information under (B) — "head bar wider, more than width of eye, etc." — and the accompanying illustrations of *A. melanopus* and *A. mccullochi* do not agree. Therefore, *A. nigripes* has been correctly identified.

## GROUP A: NO WHITE BARS

**1 BODY RED OR BLACKISH**

**A:** Pelvic and anal fins red or orange

*A. ephippium* (Andaman Sea and Java Sea)

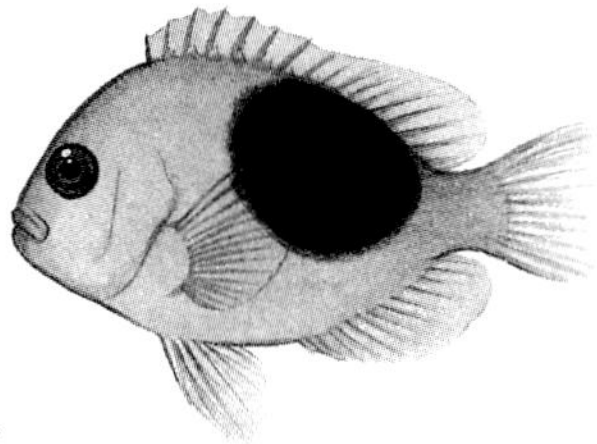

**B:** Pelvic and anal fins black

*A. melanopus* (variety from Coral Sea)

# ANEMONEFISHES

## 2 BODY ORANGE OR PINKISH

**A:** White stripe on mid-dorsal line of head extends onto lips; body bright orange

*A. sandaracinos* (western Pacific)

**B:** White stripe on mid-dorsal line of head does not extend onto lips; body generally more pinkish than orange, at least dorsally

*A. akallopisos* (Indian Ocean)

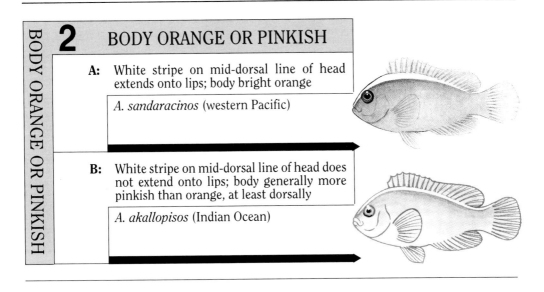

## GROUP B: ONE WHITE BAR

## 1 PELVIC AND/OR ANAL FINS DARK

**A:** Head bar narrow, equal to or less than width of eye; black colour on body restricted to breast and belly

*A. nigripes* (Maldive Islands and Sri Lanka)

**B:** Head bar wider, more than width of eye; black colour on body not restricted to breast and belly

**1:** Breast, belly, and dorsal fin red

*A. melanopus* (western Pacific)

**2:** Breast, belly, and dorsal fin dark

*A. mccullochi* (Lord Howe and Norfolk Islands)

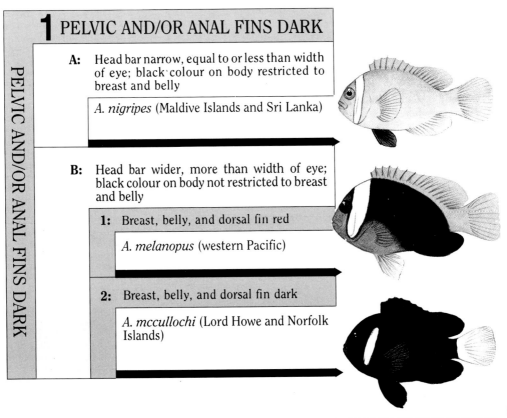

## ANEMONEFISHES

## 2 PELVIC AND ANAL FINS LIGHT (red to whitish)

**A:** Broad teardrop shaped "cap" of white on top of head; head bars usually isolated

*A. leucokranos* (Melanesia)

**B:** Head bar continuous over top of head without teardrop shaped "cap"

**1:** Head bar narrower than eye width; narow pale stripe on forehead.

*A. perideraion* (western Pacific)

**2:** Head bar usually wider than eye; no pale stripe on forehead

**A:** White saddle mark on top of tail base

*A. thiellei* (Philipines)

**B:** No white saddle on top of tail base

**1:** Head bar with well defined, but usually narrow, black margins; sides either entirely red or blackish

*A. frenatus* (South China Sea to Japan)

**2:** Head bar without well defined black margins; sides either entirely red, mainly blackish, or red with a black patch

**A:** Sides mainly blackish

*A. rubrocinctus* (northwestern Australia)

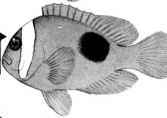

**B:** Sides with a black patch

*A. melanopus* (variety from Vanuatu - New Caledonia)

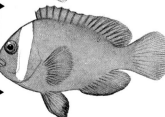

**C:** Sides entirely red

*A. melanopus* (variety from Fiji and Polynesia)

# ANEMONEFISHES

## GROUP C: TWO WHITE BARS

**(if tail bar is present, it is not well contrasted with tail colour)**

**MID-BODY BAR ABBREV. OR SLANTED**

### 1 MID-BODY BAR ABBREVIATED OR SLANTED

**A:** Dark wedge-shaped or triangular marking on tail

*A. polymnus* (Indo-Australian Archipelago)

**B:** Tail plain yellow, without dark marking

*A. sebae* (northern Indian Ocean)

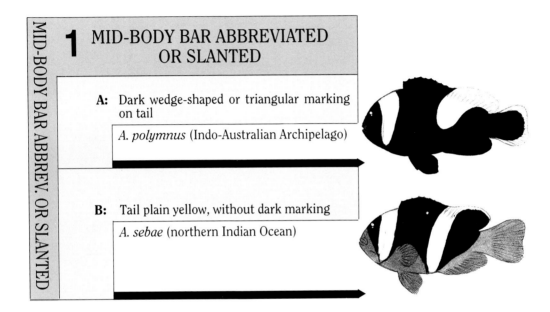

**MID-BODY BAR ALWAYS FULL**

### 2 MID-BODY BAR ALWAYS FULL

**A:** Dorsal fin dark

**1:** Tail dark

*A. tricinctus* (variety from Marshall Islands)

**2:** Tail light (white or yellow)

*A. clarkii* (Indo-West Pacific)

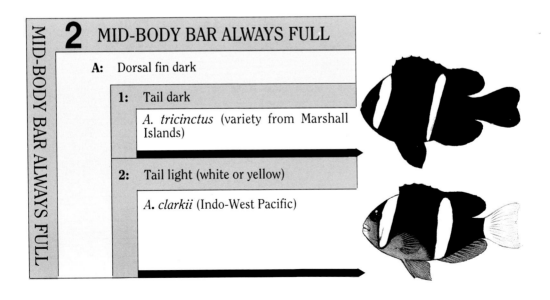

# II ANEMONEFISHES

## 2 MID-BODY BAR ALWAYS FULL

**B:** Dorsal fin lighter (yellow, orange, or light brown)

**1:** Tail distinctly forked

   **A:** Mid-body bar broad, more than 9 scales wide

      *A. latifasciatus* (Madagascar and Comoro Islands)

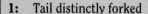

   **B:** Mid-body bar narrow, usually less than 5 scales wide

      *A. omanensis* (Oman)

**2:** Tail slightly emarginate or nearly straight

   **A:** Tail white

      **1:** Pelvic and anal fins yellow to dusky brown (not black); body mainly light to medium brown

         **A:** Pelvic fins dusky or medium brown

            *A. chagosensis* (Chagos Archipelago)

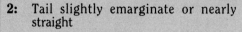

         **B:** Pelvic fins yellow to light brown

            **1:** Mainly Oceania (islands of Melanesia, Micronesia, and Polynesia)

                *A. chrysopterus* (light variety)

            **2:** Mainly eastern Australia - Coral Sea

                *A. akindynos*

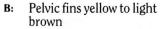

# ANEMONEFISHES II

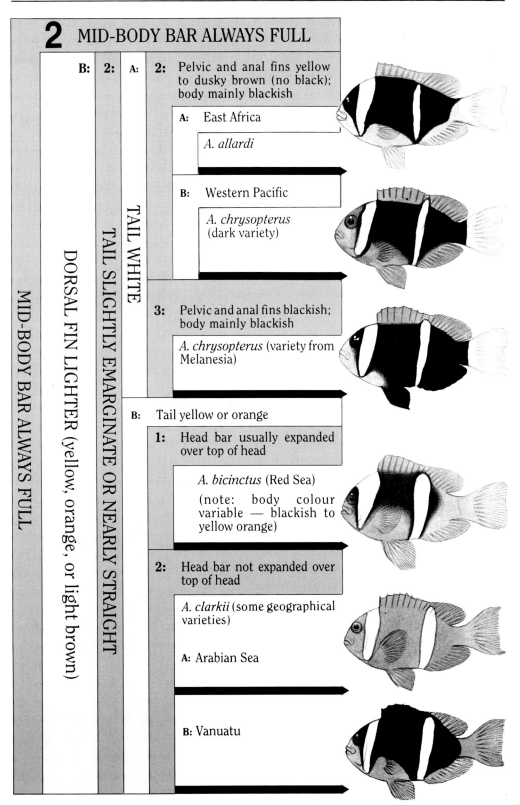

**2  MID-BODY BAR ALWAYS FULL**

**MID-BODY BAR ALWAYS FULL**

**DORSAL FIN LIGHTER** (yellow, orange, or light brown)

**B: 2: TAIL SLIGHTLY EMARGINATE OR NEARLY STRAIGHT**

**A: TAIL WHITE**

**2:** Pelvic and anal fins yellow to dusky brown (no black); body mainly blackish

- **A:** East Africa
  *A. allardi*

- **B:** Western Pacific
  *A. chrysopterus* (dark variety)

**3:** Pelvic and anal fins blackish; body mainly blackish

*A. chrysopterus* (variety from Melanesia)

**B:** Tail yellow or orange

**1:** Head bar usually expanded over top of head

*A. bicinctus* (Red Sea)
(note: body colour variable — blackish to yellow orange)

**2:** Head bar not expanded over top of head

*A. clarkii* (some geographical varieties)

**A:** Arabian Sea

**B:** Vanuatu

**II**                  ANEMONEFISHES

# GROUP D: THREE CLEARLY DEFINED BARS

### (tail bar well contrasted, i.e. tail much darker than bar)

**MID-BODY BAR WITH PRONOUNCED FORWARD BULGE**

**1 MID-BODY BAR WITH PRONOUNCED FORWARD BULGE**

- **A:** Body entirely blackish (except for bars)

  *A. ocellaris* (variety from vicinity of Darwin, Australia)

- **B:** Body at least partly orange

  - **1:** White bars with distinct black margins that are sometimes greatly expanded

    *A. percula* (melanistic variety from Melanesia)

  - **2:** White bars with thin, indistinct black margins

    - **A:** Dorsal spines usually 11 in number; height of spinous dorsal fin 3.1 to 3.3 in length of head

      *A. ocellaris* (Indo-Malayan Archipelago to Japan)

    - **B:** Dorsal spines usually 9 or 10 in number; height of spinous dorsal fin 2.1 to 2.9 in length of head

      *A. percula* (Melanesia and Queensland)

# ANEMONEFISHES

## 2 — MID-BODY BAR WITHOUT FORWARD BULGE

**A:** Body mainly red; enlarged spine(s) on cheek

*Premnas biaculeatus* (Indo-Australian Archipelago)

**B:** Body mainly dark brown to blackish; no enlarged spines on cheek

**1:** Mid-body bar broader than dark spaces between bars

*A. latezonatus* (Lord Howe and Norfolk Islands)

**2:** Mid-body bar much narrower than dark spaces between bars

**A:** Tail bar very narrow (less than 3 scales wide); breast and belly orange

*A. tricinctus* (Marshall Islands)

**B:** Tail bar wider (more than 3 scales wide); breast and belly yellow

**1:** Tail blackish on basal half with dark longitudinal streaks radiating posteriorly

*A. fuscocaudatus* (Seychelles Islands)

**2:** Tail uniformly blackish

*A. chrysogaster* (Mauritius)

# *Amphiprion akallopisos*
## BLEEKER, 1853
### SKUNK ANEMONEFISH

*Original description*:

As *Amphiprion akallopisos*, from specimens collected on Sumatra (an island of Indonesia).

*Colour features and size*:

Pink to nearly orange; without crossbars, but with a relatively narrow white stripe from top of the head to beginning of the dorsal fin, and continuing along base of the fin its entire length. Maximum length 100-110 mm.

*Similar species*:

*Amphiprion sandaracinos* is very similar, but brighter orange in colour, and its more vivid white stripe is broader; teeth of *A. sandaracinos* are conical rather than incisiform as in *A. akallopisos*.

*Distribution*:

Widespread in Indian Ocean, including Madagascar, Comoro Islands, Seychelles, Andaman Islands, west coast of Thailand, and western and southern coasts of Sumatra and Java. It also occurs in the Java Sea.

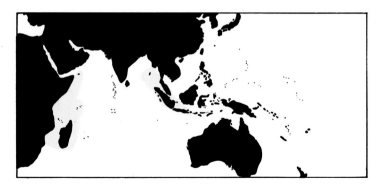

*Host anemone species*:

Heteractis
magnifica

Stichodactyla
mertensii

▲ *A. akallopisos* with *H. magnifica*, Seychelles Islands. (G. Allen)

▲ *A. akallopisos*, subadult, Phuket, Thailand. (G. Allen)

# *Amphiprion akindynos*
## ALLEN, 1972
## BARRIER REEF ANEMONEFISH

*Original description*:

As *Amphiprion akindynos* from specimens collected at the Capricorn Group, Great Barrier Reef.

*Colour features and size*:

Light to dark brown with two white bars and whitish caudal fin; head bar of adult usually constricted or discontinuous on top of head; transition between dark colour of body and pale tail not abrupt (as in *A. clarkii*), at least in adults; many juveniles and subadults have pronounced white saddle or wedge-shaped mark on upper part of tail base. Maximum length 120-130 mm.

*Similar species*:

*Amphiprion clarkii* generally has broader white bars and a very abrupt transition between the dark colour of the body and pale tail. The head bar of *A. chrysopterus* is broader and never constricted or discontinuous across its top; large adults are darker, have more orange colour, and their bars tend to be bluish.

*Distribution*:

Great Barrier Reef of Australia and adjacent Coral Sea to New Caledonia and the Loyalty Islands.

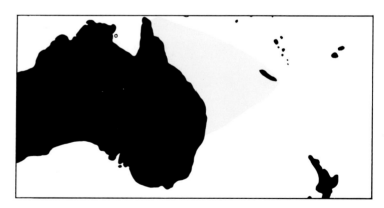

*Host anemone species*:

*Entacmaea quadricolor*

*Heteractis aurora*

*Heteractis crispa*

*Heteractis magnifica*

*Stichodactyla haddoni*

*Stichodactyla mertensii*

A. *akindynos* with *H. crispa*, Great Barrier Reef. (G. Allen)

A. *akindynos* with *E. quadricolor*, Great Barrier Reef. (G. Allen)

# *Amphiprion allardi*
## KLAUSEWITZ, 1970
## ALLARD'S ANEMONEFISH

*Original description:*

As *Amphiprion allardi*, from specimens collected at Mombasa, Kenya.

*Colour features and size:*

Dark brown to black with two white or bluish-white bars; caudal fin pale, all other fins orange. Maximum length about 140 mm.

*Similar species:*

*Amphiprion chrysopterus*, from the western Pacific Ocean, is nearly identical in colour, but the wide geographic separation between these species is sufficient to prevent confusion. Of species with ranges nearer that of *A. allardi*, *A. latifasciatus* of Madagascar and the Comoro Islands is similar, but has a wider mid-lateral bar and its yellow caudal fin is forked.

*Distribution:*

East Africa between Kenya and Durban.

*Host anemone species:*

*Entacmaea quadricolor*   *Heteractis aurora*   *Stichodactyla mertensii*

# ANEMONEFISHES    II

▲ *A. allardi* with *S. mertensii*, Kenya. (G. Allen)

▲ *A. allardi* with *E. quadricolor*, Kenya. (G. Allen)

# *Amphiprion bicinctus*
## RÜPPELL, 1828
## TWO-BAND ANEMONEFISH

*Original description*:

As *Amphiprion bicinctus*, from specimens collected at Massaua, Red Sea.

*Colour features and size*:

Bright orange to dark brown with two white or bluish-white bars, the first considerably expanded (rarely narrow) across top of the head. Maximum length about 140 mm.

*Similar species*:

Many species have a similar pattern of two white bars, but *A. bicinctus* differs from nearly all of them by having a yellowish caudal fin (it is whitish in other species); and the expansion of the first bar over the top of the head differs from the narrower bar typical of most other species (although the bar is occasionally narrow in *A. bicinctus*). *Amphiprion latifasciatus* from Madagascar and the Comoro Islands is similar in pattern and also has a yellow tail, but the mid-body bar is much wider and the tail is forked (it is truncate or only slightly emarginate in *A. bicinctus*). *Amphiprion chagosensis* from the Chagos Archipelago and *A. allardi* from eastern Africa have a white tail.

*Distribution*:

Red Sea, Gulf of Aden, and Chagos Archipelago.

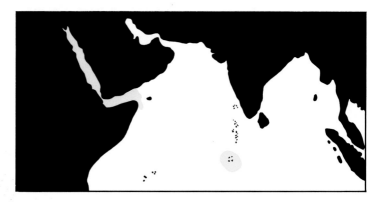

*Host anemone species*:

*Entacmaea quadricolor*

*Heteractis aurora*

*Heteractis crispa*

*Heteractis magnifica*

*Stichodactyla gigantea*

*Stichodactyla mertensii*

▲ *A. bicinctus*, Jeddah, Red Sea. (G. Allen)

▲ *A. bicinctus* with *H. magnifica*, Jeddah, Red Sea. (G. Allen)

# Amphiprion chagosensis
## ALLEN, 1972
## CHAGOS ANEMONEFISH

*Original description*:

As *Amphiprion chagosensis*, from specimens collected at Diego Garcia Atoll, Chagos Archipelago.

*Colour features and size*:

Light to dark brown with two dark-edged white bars; dorsal, anal, and pelvic fins of adult dusky brown; caudal fin whitish. Maximum length about 100 mm.

*Similar species*:

*Amphiprion akindynos* from the Great Barrier Reef-Coral Sea region is similar in appearance, but has paler dorsal, anal, and pelvic fins (at least in adults), and the first white bar is usually constricted or interrupted on top of the head, In addition, *A. chagosensis* differs from similar two-barred species in the extent of predorsal scalation: the top of its head is scaled to a point nearly even with the front of the eyes, whereas in *A. akindynos, A. allardi*, and *A. bicinctus*, scales extend only to the rear of the eyes or, at most, are even with the middle of the eyes.

*Distribution*:

Chagos Archipelago in the central Indian Ocean. Recently observed at Sharm-el-Sheikh, Red Sea.

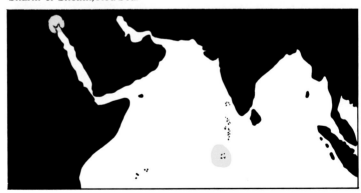

*Host anemone species*:

*Entacmaea quadricolor.*

# ANEMONEFISHES

▲ *A. chagosensis*, Sharm-el-Sheikh, Red Sea. (J. Kemp)

▲ *A. chagosensis*, subadult, Diego Garcia Atoll, central Indian Ocean. (R. Winterbottom)

# *Amphiprion chrysogaster*
## CUVIER, 1830
## MAURITIAN ANEMONEFISH

*Original description*:

As *Amphiprion chrysogaster*, from specimens collected at Mauritius (southwestern Indian Ocean).

*Colour features and size*:

Very dark brown (nearly black) with three white bars; breast and snout orange; anal fin either orange-yellow or blackish; caudal fin dark brown or blackish. Maximum length about 140 mm.

*Similar species*:

*Amphiprion fuscocaudatus* from the Seychelles has a similar colour pattern but its tail has dark streaks rather than being solid dark colour and it has 11 dorsal spines (there are 10 in *A. chrysogaster*). *Amphiprion tricinctus*, from the Marshall Islands in the Pacific, is also similar, but generally has a narrower bar across the tail base (less than three scales wide — it is more than three scales wide in *A. chrysogaster*).

*Melanistic variation*:

Specimens associated with *Stichodactyla mertensii* are generally blackish except for the three white bars.

*Distribution*:

Mauritius and probably Réunion.

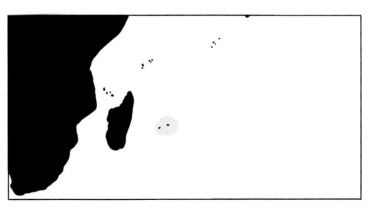

*Host anemone species*:

*Heteractis aurora*  *Heteractis magnifica*  *Macrodactyla doreensis*  *Stichodactyla haddoni*  *Stichodactyla mertensii*

▲ *A. chrysogaster* with *H. magnifica*, Mauritius. (J. Post/Ikan)

▲ *A. chrysogaster* with *H. magnifica*, Mauritius. (G. Allen)

# ANEMONEFISHES

## *Amphiprion chrysopterus*
### CUVIER, 1830
### ORANGE-FIN ANEMONEFISH

*Original description*:

As *Amphiprion chrysopterus*, from specimens of undetermined origin.

*Colour features and size*:

Brown to nearly black with two white or bluish-white bars and a whitish caudal fin; over most of the western Pacific all other fins are yellow-orange, but fish from Melanesia have black pelvic and anal fins. Maximum length about 150 mm.

*Similar species*:

Three species with overlapping distributions may be confused with *A. chrysopterus*. Adults of *A. akindynos* from the Great Barrier Reef - Coral Sea region tend to be lighter brown with pelvic and anal fins that are never black; *Amphiprion clarkii* has a wider mid-lateral bar and almost always has a third bar across the tail base; *A. tricinctus* has a bar across the tail base except when solid black, in which case it has only two bars.

*Melanistic variation*:

Fish living with *Stichodactyla mertensii* generally have a blackish ground colour, whereas males and juveniles that occupy *Heteractis crispa* are brown. Only orange or brown juveniles are found with *H. aurora*.

*Distribution*:

Widespread in the western Pacific including New Guinea, Coral Sea, New Britain, Solomon Islands, Vanuatu, Fiji, Caroline Islands, Mariana Islands, Gilbert Islands, Samoa, Society Islands, and Tuamotu Islands.

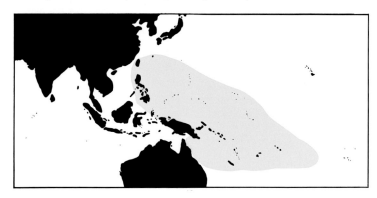

*Host anemone species*:

*Entacmaea quadricolor*

*Heteractis aurora*

*Heteractis crispa*

*Heteractis magnifica*

*Stichodactyla haddoni*

*Stichodactyla mertensii*

## ANEMONEFISHES  II

▲ *A. chrysopterus* with *S. mertensii*, Micronesia. (E. Robinson/Ikan)

▲ *A. chrysopterus*, Manus Island, Papua New Guinea. (G. Allen)

## ANEMONEFISHES

# *Amphiprion clarkii*
## (BENNETT, 1830)
### CLARK'S ANEMONEFISH

*Original description*:

As *Anthias clarkii*, from specimens collected at Ceylon (now Sri Lanka).

*Colour features and size*:

Usually black with variable amount of orange on head, ventral parts, and fins; three white bars — on head, body, base of caudal fin; transition between darker body and bar across caudal fin base usually abrupt; caudal fin usually white or white with yellow edges (males), but sometimes yellow; juveniles from all areas and adults from Vanuatu and New Caledonia may be mostly or entirely orange-yellow with only two anterior white bars. Maximum length about 140 mm.

*Similar species*:

*Amphiprion latifasciatus* (Madagascar and Comoro Islands) lacks a white bar on the caudal fin base and its tail is forked. *Amphiprion allardi* (East Africa), *A. akindynos* (Great Barrier Reef - Coral Sea), and *A. chagosensis* have a narrower mid-body bar and lack the sharp demarcation between white on the caudal fin base and dark of the body. *Amphiprion chrysogaster* (Mauritius), *A. fuscocaudatus* (Seychelles), and *A. tricinctus* (Marshall Islands) have three white bars, but the caudal fin is dark.

*Melanistic variation*:

Fish that live with *Stichodactyla mertensii* are frequently black except for pale snout, white bars, and yellow or white tail.

*Distribution*:

*Amphiprion clarkii* is the most widely distributed anemonefish, ranging from the islands of Micronesia and Melanesia in the western Pacific to the Persian Gulf, and from Australia to Japan.

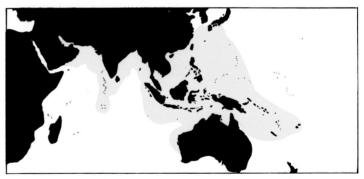

*Host anemone species*:

 *Cryptodendrum adhaesivum*

 *Entacmaea quadricolor*

 *Heteractis aurora*

 *Heteractis crispa*

 *Heteractis magnifica*

 *Heteractis malu*

 *Macrodactyla doreensis*

 *Stichodactyla gigantea*

 *Stichodactyla haddoni*

 *Stichodactyla mertensii*

# ANEMONEFISHES  II

▲ *A. clarkii*, Madang, Papua New Guinea. (G. Allen)

▲ *A. clarkii* with *H. crispa*, Madang, Papua New Guinea. (G. Allen)

## *Amphiprion ephippium*
(BLOCH, 1790)
### RED SADDLEBACK ANEMONEFISH

*Original description*:

As *Lutjanus ephippium*, from specimens collected at Tranquebar (about 250 km south of Madras, India; probably an erroneous locality as it is outside the known range).

*Colour features and size*:

Body and fins reddish-orange, lacking white bars (except small juveniles); black spot or saddle on sides usually varies in size with size of individual, being small in subadults and covering much of the posterior body in mature fish. Maximum length about 120 mm.

*Similar species*:

*Amphiprion frenatus* (South China Sea to Japan), *A. melanopus* (western Pacific), and *A. rubrocinctus* (northwestern Australia) are somewhat similar, but all possess a single white head bar as adults (juveniles may have 2-3 bars); in addition, scales on top of the head in these species do not extend as far forward (to about the middle part of the eyes) as in *A. ephippium* (to a level even with the front of the eyes).

*Distribution*:

Andaman and Nicobar Islands, Thailand, Malaysia, Sumatra, and Java.

*Host anemone species*:

*Entacmaea quadricolor*

*Heteractis crispa*

▲ *A. ephippium*, Indonesia. (H. Debelius/ Ikan)

▲ *A. ephippium*, subadult, with *H. crispa*, Java Sea. (G. Allen)

# *Amphiprion frenatus*
## BREVOORT, 1856
### TOMATO ANEMONEFISH

*Original description*:

As *Amphiprion frenatus*, from specimens collected at Japan.

*Colour features and size*:

Adults with a single white head bar; females mainly blackish on sides with red snout, breast, belly, and fins; males considerably smaller than females and lacking blackish colouration — being instead red overall; juveniles with two or three white bars. Maximum length about 140 mm.

*Similar species*:

*Amphiprion rubrocinctus* (northwestern Australia) is very similar in colouration, but its white bar lacks the distinctive black outline of *A. frenatus*; the bar in females is poorly developed, has an irregular outline, and is sometimes discontinuous on top of the head; the smaller male generally has blackish sides. Small juveniles of *A. frenatus* and *A. rubrocinctus* are very difficult to separate; because they do not have overlapping distributions, geography is the best means of distinguishing them. *Amphiprion melanopus* (western Pacific) is also similar, but generally has a broader white head bar, and specimens from most areas (except eastern Melanesia) have black pelvic and anal fins.

*Melanistic variation*:

Only that related to sex as described above.

*Distribution*:

South China Sea and immediately adjacent areas, northwards to Japan.

*Host anemone*:

*Entacmaea quadricolor*

## ANEMONEFISHES II

▲ *A. frenatus* with *E. quadricolor*, Okinawa, Japan. (G. Allen)

▲ *A. frenatus* with *E. quadricolor*, Sabah, Borneo. (G. Allen)

# *Amphiprion fuscocaudatus*
## ALLEN, 1972
## SEYCHELLES ANEMONEFISH

*Original description*:

As *Amphiprion fuscocaudatus*, from specimens collected at the Seychelles Islands (northwestern Indian Ocean).

*Colour features and size*:

Dark brown to blackish with three white bars; snout, breast, belly, and pelvic and anal fins yellow-orange; dorsal and caudal fins dusky brown to blackish. Maximum length about 140 mm.

*Similar species*:

*Amphiprion chrysogaster* (Mauritius) is very similar in having three white bars and a dark caudal fin. However, the caudal fin of *A. chrysogaster* is uniformly dark except for a narrow white margin, whereas in *A. fuscocaudatus*, it has a dark central area at its base, with dark longitudinal streaks separated by lighter areas radiating from it.

*Distribution*:

Seychelles Islands and Aldabra in the western Indian Ocean.

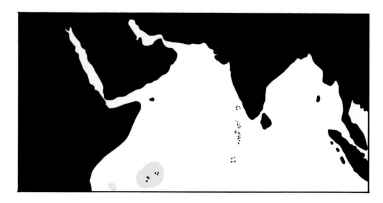

*Host anemone species*:

*Stichodactyla mertensii*

# ANEMONEFISHES

▲ *A. fuscocaudatus* with *S. mertensii*, Seychelles Islands. (J. Randall)

▲ *A. fuscocaudatus*, Seychelles Islands. (G. Allen)

# *Amphiprion latezonatus*
## WAITE, 1900
### WIDE-BAND ANEMONEFISH

*Original description*:

As *Amphiprion latezonatus*, from specimens collected at Lord Howe Island (southwestern Pacific Ocean).

*Colour features and size*:

Dark brown with three white bars; mid-body bar extremely wide and shaped like flat-topped pyramid; caudal fin dark brown with broad, pale posterior margin. Maximum length about 140 mm.

*Similar species*:

One of the most distinctive anemonefishes, *Amphiprion latezonatus* is unlikely to be confused with any other. Its mid-body bar is more than twice the width of this bar in most other species.

*Distribution*:

Lord Howe Island off eastern Australia and rocky mainland reefs near the Queensland - New South Wales border.

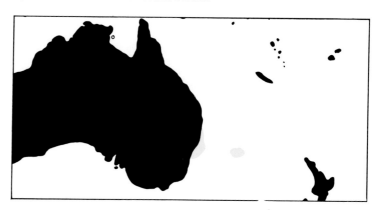

*Host anemone species*:

*Heteractis crispa*

# ANEMONEFISHES

▲ *A. latezonatus*, New South Wales, Australia. (G. Allen)

▲ *A. latezonatus* with *H. crispa*, New South Wales, Australia. (G. Allen)

# *Amphiprion latifasciatus*
## ALLEN, 1972
## MADAGASCAR ANEMONEFISH

*Original description*:

As *Amphiprion latifasciatus*, from specimens collected at Madagascar.

*Colour features and size*:

Blackish with two white bars; snout, belly, and all fins, including tail, yellow; caudal fin slightly forked. Maximum length about 130 mm.

*Similar species*:

*Amphiprion bicinctus* (Red Sea), *A. allardi* (East Africa), *A. chrysopterus* (western Pacific), and *A. clarkii* (Indo-West Pacific) have a similar colour pattern. The mid-body bar of *A. latifasciatus* is generally wider than in these species, and its caudal fin is forked rather than truncate to slightly emarginate. It further differs from *A. chrysopterus* and *A. allardi* (and most individuals of *A. clarkii*) in having a yellow rather than a white tail, and from most *A. clarkii* in lacking a white bar at the base of the caudal fin. *Amphiprion omanensis* (Arabian Sea) also has a forked caudal fin, but the midbody bar is much narrower (1½ to 4 scales wide) and the pelvic and anal fins are black.

*Distribution*:

Madagascar and the Comoro Islands in the western Indian Ocean.

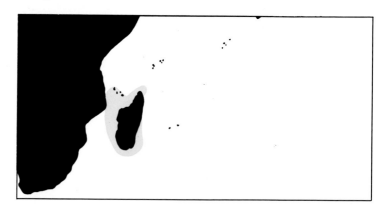

*Host anemone species*:

*Stichodactyla mertensii*

▲ *A. latifasciatus* with *S. mertensii*, Comoro Islands. (M. Parmantier, courtesy of Nancy Aquarium, France)

▲ A rare hybrid between *A. leucokranos* and *A. chrysopterus*, Madang, Papua New Guinea. (G. Allen). Refer to text on page 49.

# *Amphiprion leucokranos*
## ALLEN, 1973
## WHITE-BONNET ANEMONEFISH

*Original description*:

As *Amphiprion leucokranos*, from specimens collected at Madang, Papua New Guinea.

*Colour features and size*:

Orange to light brown with a large teardrop shaped white area on top of head and a single white bar on each side of head that may or may not be connected to the white area above it. Maximum length about 90 mm.

*Similar species*:

*Amphiprion sandaracinos* is a similar colour, but lacks the white head bar and mark on top of the head, having, instead, a white mid-dorsal stripe extending from the snout along the spine to the base of the tail.

*Distribution*:

Northern Papua New Guinea, including Manus Island and New Britain, and the Solomon Islands.

*Host anemone species*:

*Heteractis crispa*

*Heteractis magnifica*

*Stichodactyla mertensii*

*Remarks*:

This "species" is probably a hybrid between *A. chrysopterus* and *A. sandaracinos.*

*A. leucokranos* with *H. magnifica*, Solomon Islands. (R. Steene)

*A. leucokranos* with *H. crispa*, Madang, Papua New Guinea. (G. Allen)

# *Amphiprion mccullochi*
## WHITLEY, 1929
## McCULLOCH'S ANEMONEFISH

*Original description*:

As *Amphiprion mccullochi*, from specimens collected at Lord Howe Island (southwestern Pacific Ocean).

*Colour features and size*:

Dark brown with whitish snout and caudal fin; white bar on each side of head usually not connected on top of head in adults. Maximum length about 120 mm.

*Similar species*:

*Amphiprion melanopus* (western Pacific) is similar, but has reddish coloured breast, belly, and dorsal fin, and the caudal fin is yellowish to slightly red. In addition, the white bars are interconnected over the top of the head.

*Distribution*:

Lord Howe Island off New South Wales, Australia, and nearby Norfolk Island.

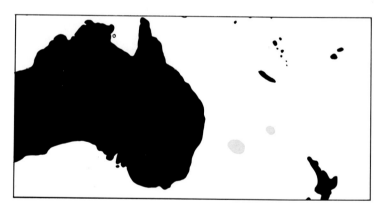

*Host anemone species*:

*Entacmaea quadricolor*

## ANEMONEFISHES                                                    II

▲ *A. mccullochi*, with *E. quadricolor*, Lord Howe Island, Australia. (J. Randall)

▲ *A. mccullochi*, Lord Howe Island, Australia. (W. Doak)

# *Amphiprion melanopus*
## BLEEKER, 1852
## RED AND BLACK ANEMONEFISH

*Original description*:

As *Amphiprion melanopus*, from specimens collected at Ambon (Molucca Islands, Indonesia).

*Colour features and size*:

Adults usually black on sides with reddish snout, belly, dorsal fin, and tail (sometimes pale yellow); pelvic and anal fins usually black; a single relatively broad white bar on head. Some individuals from the Coral Sea lack head bar; fish from the Fiji Islands and southeastern Polynesia entirely red except for white head bar; those from the Solomon Islands, Vanuatu, and New Caledonia have reduced black patch on the side. Maximum length about 120 mm.

*Similar species*:

In the normal adult colouration, the black pelvic and anal fins easily distinguish *A. melanopus* from the other single-barred, red-finned species, *A. frenatus* (South China Sea to Japan) and *A. rubrocinctus* (northwestern Australia). However, Fijian and southeastern Polynesian specimens are readily confused with the red males of *A. frenatus*. The best means of separation is the pronounced black border on the margins (particularly the rear one) of the white head bar in *A. frenatus*, which is lacking in *A. melanopus*.

*Melanistic variation*:

None except variation between "normal" dark colour phase and red "Fijian" phase noted above.

*Distribution*:

Indonesia (Bali eastward), Melanesia, Micronesia, southeastern Polynesia, and Great Barrier Reef - Coral Sea.

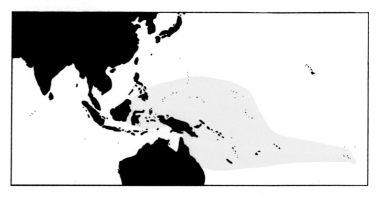

*Host anemone species*:

Usually *Entacmaea quadricolor*     Occasionally *Heteractis crispa*     Rarely *Heteractis magnifica*

▲ *A. melanopus* with *E. quadricolor*, Great Barrier Reef. (G. Allen)

▲ Variant of *A. melanopus* with *E. quadricolor*, Holmes Reef, Coral Sea. (G. Allen)

# *Amphiprion nigripes*
## REGAN, 1908
## MALDIVES ANEMONEFISH

*Original description*:

As *Amphiprion nigripes*, from specimens collected at the Maldive Islands.

*Colour features and size*:

Pale orange brown with narrow white head bar; belly, pelvic fins, and anal fin black. Maximum length about 110 mm.

*Similar species*:

No other species in the Indian Ocean except *A. perideraion* (eastern Indian and western Pacific Oceans) has only one white bar; *A. nigripes* is easily separated from it and all other anemonefishes by the combination of the single white bar and black belly, pelvic fins, and anal fins.

*Distribution*:

Maldive Islands and Sri Lanka in the central Indian Ocean.

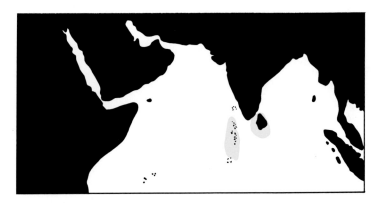

*Host anemone species*:

*Heteractis magnifica*

# ANEMONEFISHES                                                              II

▲ *A. nigripes* with *H. magnifica*, Maldive Islands. (G. Allen)

▲ *A. nigripes* with *H. magnifica*, Maldive Islands. (H. Debelius/Ikan)

# *Amphiprion ocellaris*
## CUVIER, 1830
## FALSE CLOWN ANEMONEFISH

*Original description*:

As *Amphiprion ocellaris*, from specimens collected at Sumatra (an island of Indonesia).

*Colour features and size*:

Normally bright orange with three white bars, the middle one with forward-projecting bulge; bars have narrow black margins. Maximum length about 90 mm.

*Similar species*:

*Amphiprion percula* (northern Queensland and Melanesia) is nearly identical, but has 10 (rarely 9) dorsal spines compared to 11 (rarely 10) in *A. ocellaris*; the spinous (anterior) part of the dorsal fin in *A. ocellaris* is taller (its height fits about 2.1-2.9 in the head length compared to 3.1-3.3 in *A. percula*). Distributions of these two species do *not* overlap.

*Melanistic variation*:

A variety that is entirely black except for the white bars occurs in the vicinity of Darwin, Australia. Whether this melanism is correlated with a particular species of anemone is uncertain.

*Distribution*:

Andaman and Nicobar Islands (Andaman Sea), Indo-Malayan Archipelago, Philippines, northwestern Australia, coast of Southeast Asia northwards to the Ryukyu Islands.

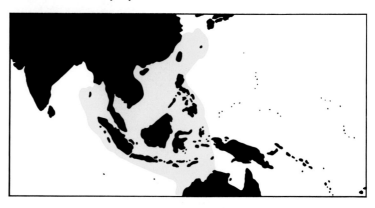

*Host anemone species*:

*Heteractis magnifica*

*Stichodactyla gigantea*

*Stichodactyla mertensii*

▲ *A. ocellaris* with *H. magnifica*, Philippine Islands. (H. Hall/Ikan)

▲ Unusual dark variety of *A. ocellaris*, Sipadan Island, Borneo. (G. Allen)

# *Amphiprion omanensis*
## ALLEN AND MEE, 1991
## OMAN ANEMONEFISH

*Original Description:*

As *Amphiprion omanensis*, from specimens collected at Oman (described in Allen, 1991).

*Colour features and size:*

Body medium to dark brown, head lighter (very pale tannish on snout and chin); two white bars, the head bar usually constricted on forehead, midbody bar narrow, about 1½-4 scales wide; dorsal fin brown to tan; caudal fin tan to whitish, pelvic and anal fins black, pectoral fins yellowish.

*Similar species:*

The combination of a strongly forked caudal fin and black pelvic and anal fins is distinctive. The only other species with a forked caudal fin is *A. latifasciatus* from Madagascar and the Comoro Islands. However, it has a much wider midbody bar, usually about 10 scales wide.

*Distribution:*

Oman, Arabian Peninsula.

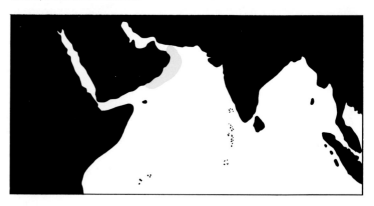

*Host anemone species:*

Entacmaea quadricolor

Heteractis crispa

Stichodactyla haddoni

## ANEMONEFISHES

▲ *A. omanensis*, Arabian Sea. (H. Debelius/Ikan)

▲ *A. omanensis* with *E. quadricolor*, Arabian Sea. (H. Debelius/Ikan)

# *Amphiprion percula*
## (LACEPÈDE, 1802)
## CLOWN ANEMONEFISH

*Original description*:

As *Lutjanus percula*, from specimens collected at New Britain (now part of Papua New Guinea).

*Colour features and size*:

Bright orange with three white bars, the middle with forward-projecting bulge; bars often bordered with black that varies in width. Maximum length about 80 mm.

*Similar species*:

*Amphiprion ocellaris* is nearly identical, but has 11 (rarely 10) dorsal spines compared to 10 (rarely 9) in *A. percula*; the spinous (anterior) part of the dorsal fin of *A. ocellaris* is taller (its height fits about 2.1-2.9 in the head length compared to 3.1-3.3 in *A. percula*); *A. ocellaris* never has a thick black margin around the white bars. These two species do *not* have overlapping distributions.

*Melanistic variation*:

Limited melanism is evident in fish that live with anemones of the genus *Stichodactyla*: the margin around the white bars is deep black, and, in some specimens, considerably expanded.

*Distribution*:

Northern Queensland and Melanesia (New Guinea, New Britain, New Ireland, Solomon Islands, and Vanuatu).

*Host anemone species*:

*Heteractis crispa*  *Heteractis magnifica*  *Stichodactyla gigantea*

## ANEMONEFISHES

▲ *A. percula* with *S. gigantea*, Hermit Islands, Papua New Guinea. (E. Robinson/Ikan)

▲ *A. percula* with *S. gigantea*, Madang, Papua New Guinea. (G. Allen)

# *Amphiprion perideraion*
## BLEEKER, 1855
## PINK ANEMONEFISH

*Original description*:

As *Amphiprion perideraion*, from specimens collected at Groot Oby (Obi Island, Molucca Islands, Indonesia).

*Colour features and size*:

Pink to pinkish orange; fins pale; narrow white head bar and white stripe on top of head beginning between the eyes and extending along base of dorsal fin; adult males have narrow orange margin on soft dorsal fin and upper and lower edges of tail. Maximum length about 100 mm.

*Similar species*:

*Amphiprion nigripes* (Maldives and Sri Lanka) is more reddish and has a black belly and black pelvic and anal fins; *A. leucokranos* (Melanesia) has a wider head bar and the much broader white stripe (usually teardrop shape) on top of head does not extend the full length of the dorsal fin base; *A. akallopisos* (Indian Ocean) and *A. sandaracinos* (eastern Indian - western Pacific Oceans) lack the white head bar.

*Distribution*:

Cocos (Keeling) Islands and Christmas Island in the eastern Indian Ocean, Indo-Australian Archipelago northwards to the Ryukyu Islands, and Micronesia.

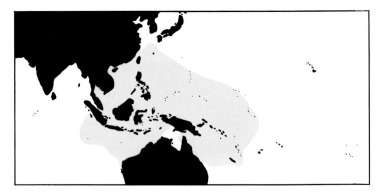

*Host anemone species*:

| *Heteractis crispa* | Usually *Heteractis magnifica* | *Macrodactyla doreensis* | *Stichodactyla gigantea* |

▲ *A. perideraion* with *H. magnifica*, Great Barrier Reef. (G. Allen)

▲ *A. perideraion* and nest of eggs with *H. magnifica*, Marshall Islands. (G. Allen)

# *Amphiprion polymnus*
(LINNAEUS, 1758)
## SADDLEBACK ANEMONEFISH

*Original description*:

As *Perca polymnus*, from specimens collected in the Indies (i.e. East Indies, now Indonesia).

*Colour features and size*:

Dark brown with broad white bar just behind eye; on middle of back an abbreviated white saddle or on middle of the side a partial to complete white bar that slants slightly backwards, extending onto middle and rear parts of dorsal fin; caudal fin mainly dark brown with broad whitish margins, the dark part tapering in width posteriorly; breast and belly either yellow-orange or dark brown. Maximum length about 120 mm.

*Similar species*:

*Amphiprion sebae* (Indian Ocean) has a yellow tail.

*Melanistic variation*:

Fish associated with *Heteractis crispa* are usually entirely dark except for the white bars and caudal fin margin and tannish snout.

*Distribution*:

Indo-Malayan Archipelago northwards to the Ryukyu Islands; also reported from the Northern Territory, Australia.

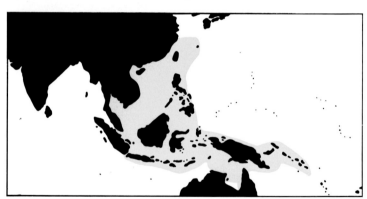

*Host anemone species*:

Heteractis crispa | Macrodactyla doreensis | Stichodactyla haddoni

## ANEMONEFISHES  II

▲ *A. polymnus* with *S. haddoni*, Madang, Papua New Guinea. (G. Allen)

▲ *A. polymnus* with *S. haddoni*, Indonesia. (R. Kuiter)

# *Amphiprion rubrocinctus*
## RICHARDSON, 1842
## AUSTRALIAN ANEMONEFISH

*Original description*:

As *Amphiprion rubrocinctus*, from specimens collected at Depuch Island, Western Australia.

*Colour features and size*:

Dark brown or blackish on sides with single pale (white to pink) bar on head; head bar often poorly developed in adults and lacking pronounced black margin; snout, breast, belly, and fins red. Maximum length about 120 mm.

*Similar species*:

*Amphiprion frenatus* (South China Sea to Japan) is similar in colour, but males of that species are entirely bright red except for the head bar, and females have a more vivid white head bar with a narrow black margin (lacking in *A. rubrocinctus*). The distributions of these species do not overlap.

*Distribution*:

Northwestern Australia.

*Host anemone species*:

Usually
*Entacmaea quadricolor*

*Stichodactyla gigantea*

# ANEMONEFISHES

A. *rubrocinctus* with *E. quadricolor*, Western Australia. (G. Allen)

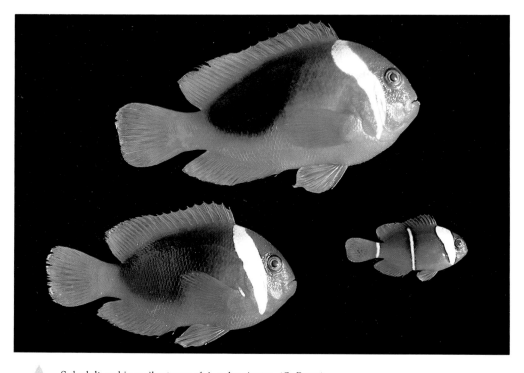

Subadult and juvenile stages of *A. rubrocinctus*. (C. Bryce)

# *Amphiprion sandaracinos*
## ALLEN, 1972
## ORANGE ANEMONEFISH

*Original description*:

As *Amphiprion sandaracinos*, from specimens collected in the Philippines.

*Colour features and size*:

Bright orange with white mid-dorsal stripe between middle of snout and upper base of tail; no white bars on head and body. Maximum length about 130 mm.

*Similar species*:

*Amphiprion akallopisos* (Indian Ocean) tends to be more pink than orange, and its mid-dorsal stripe usually does not reach the upper lip as in *A. sandaracinos*. Teeth of the two species differ in shape (flat-topped in *A. akallopisos* and conical in *A. sandaracinos*). *Amphiprion leucokranos* (Melanesia) has a similar orange colouration, but possesses a head bar and teardrop shaped mark on top of the head.

*Distribution*:

Christmas Island and Western Australia in the eastern Indian Ocean, Indonesia, Melanesia, Philippines, and northwards to the Ryukyu Islands.

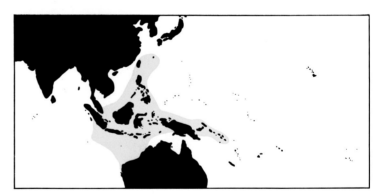

*Host anemone species*:

*Heteractis crispa*

*Stichodactyla mertensii*

## ANEMONEFISHES

▲ *A. sandaracinos* with *S. mertensii*, Madang, Papua New Guinea. (G. Allen)

▲ *A. sandaracinos* with *S. mertensii*, Christmas Island, Indian Ocean. (G. Allen)

# *Amphiprion sebae*
## BLEEKER, 1853
## SEBAE ANEMONEFISH

*Original description*:

As *Amphiprion sebae*, from specimens collected at Java (an island of Indonesia).

*Colour features and size*:

Dark brown to blackish with two white bars, the mid-body bar slanting slightly backwards and extending onto rear part of dorsal fin; snout, breast, and belly often yellow-orange; tail yellow or orange. Maximum length about 140 mm.

*Similar species*:

*Amphiprion polymnus* (Indo-Australian Archipelago to Japan) is similar, but has a distinctive dark, wedge-shaped mark covering most of the tail.

*Melanistic variation*:

Some individuals are entirely dark brown to blackish on the body (except for white bars), lacking yellow-orange colour on the snout, breast, and belly. It is not known if the variation is associated with a particular anemone host.

*Distribution*:

Northern Indian Ocean including Java, Sumatra, Andaman Islands, India, Sri Lanka, Maldive Islands, and southern Arabian Peninsula.

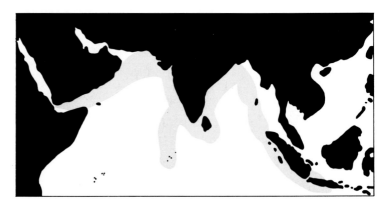

*Host anemone species*:

*Stichodactyla haddoni*

*A. sebae*, Sri Lanka. (G. Allen)

*A. sebae*, subadult, Arabian Sea. (G. Allen)

# *Amphiprion thiellei*
## BURGESS, 1981
## THIELLE'S ANEMONEFISH

*Original description:*

As *Amphiprion thiellei*, from specimens of unknown origin (via a pet shop).

*Colour features and size:*

Reddish-orange with single, relatively narrow white head bar; also small white saddle on top edge of tail base. Maximum length 65 mm.

*Similar species:*

*Amphiprion perideraion* (eastern Indian - western Pacific Oceans) is pink or pinkish-orange without the reddish hue, and has a white stripe along the base of the dorsal fin and down the middle of the forehead. *Amphiprion nigripes* (Maldive Islands and Sri Lanka) has a black belly, pelvic fins, and anal fin.

*Distribution:*

The species was described from two specimens obtained from a pet dealer; their origin is uncertain, although they are believed to have been collected in the Philippines. The species appears to be distinctive in colouration and morphology (especially a low gill-raker count), but until more specimens are studied, its status as a valid species is provisional: it might represent a rare variant of another species or a hybrid. No other specimens have been seen since the two original fish were obtained.

*Host anemone species:*

Not known

*Remarks:*

This "species" is most likely a hybrid between *A-chrysopterus* and *A-sandarcinos*.

▲ *A. thiellei*, aquarium photo. (M. Thielle)

▲ *A. thiellei*, subadult, aquarium photo. (A. Norman)

# *Amphiprion tricinctus*
## SCHULTZ AND WELANDER, 1953
## THREE-BAND ANEMONEFISH

*Original description*:

As *Amphiprion tricinctus*, from specimens collected at Bikini Atoll, Marshall Islands.

*Colour features and size*:

Black or dark brown with two or three white bars; snout, breast, belly, pelvic fins, and anal fin frequently yellow-orange; tail dark brown or black. Maximum length about 130 mm.

*Similar species*:

*Amphiprion chrysogaster* (Mauritius) and *A. fuscocaudatus* (Seychelles) both have a combination of three white bars and a dark tail, but there are broad gaps among the distributions of the three species. *Amphiprion fuscocaudatus* differs in having dark streaks radiating from the basal black area on the tail (the tail of *A. tricinctus* is uniformly dark ); *A. chrysogaster* has a wider white bar across the tail base (more than three scales wide compared to less than three in *A. tricinctus*). Both *A. chrysogaster* and *A. fuscocaudatus* tend to be more yellow on the breast, belly, and pelvic and anal fins, in contrast to the orange colour of *A. tricinctus*. The two Indian Ocean species do not possess a melanistic variety (see below).

*Melanistic variation*:

Fish associated with *Stichodactyla mertensii* are entirely black or dark brown with either two or three white bars.

*Distribution*:

Marshall Islands in the central-western Pacific Ocean.

*Host anemone species*:

*Entacmaea quadricolor*  *Heteractis aurora*  *Heteractis crispa*  *Stichodactyla mertensii*

A. tricinctus with E. quadricolor, Enewetak Atoll, Marshall Islands. (J. Randall)

A. trincinctus with E. quadricolor, Marshall Islands. (G. Allen)

# *Premnas biaculeatus*
## (BLOCH, 1790)
### SPINE-CHEEK ANEMONEFISH

*Original description*:

As *Chaetodon biaculeatus*, from specimens collected in the East Indies (now Indonesia).

*Colour features and size*:

Bright red to brownish-red with three relatively narrow white or grey bars; all fins same colour as body; cheek usually with pair of long spines. Maximum length about 160 mm. Males much smaller (usually less than 60-70 mm) and brighter red than females, with brilliant white stripes; female bars generally grey, but can be "switched" rapidly to white if fish is provoked. Fish from Sumatra possess yellow bars.

*Similar species*:

None: the cheek spines, uniformly bright red body and fin colour, and narrow white bars separate this from all other anemonefishes.

*Distribution*:

Indo-Malayan Archipelago to northern Queensland.

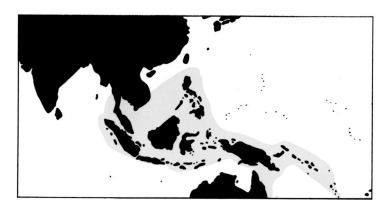

*Host anemone species*:

Entacmaea quadricolor
(usually solitary form)

▲ *P. biaculeatus* with *E. quadricolor*, Great Barrier Reef. (G. Allen)

▲ *P. biaculeatus*, Sabah, Borneo. (G. Allen)

# *Dascyllus trimaculatus*
## RÜPPELL, 1829
### THREE-SPOT DASCYLLUS

*Original description*:

As *Dascyllus trimaculatus*, from specimens collected in the Red Sea.

*Colour features and size*:

Black to light greyish (nuptial male), including fins; often with a small white spot on upper side that is larger, and therefore much more distinct, in young fish, which also have white spot on forehead. Maximum length 140 mm.

*Similar species*:

*Dascyllus albisella* (Hawaiian Islands and Johnston Island) and *D. strasburgi* (Marquesas Islands) are similar in appearance. Both are endemic to relatively small areas in which *Dascyllus trimaculatus* does not occur. The young of *D. albisella* (and possibly *D. strasburgi*) sometimes associate with the anemone *Heteractis malu*.

*Colour variation*:

Some individuals have red-orange or partly red-orange fins and adjacent parts of the body. This variation is often seen in relatively turbid water.

*Distribution*:

Widespread in the tropical Indo-West Pacific from East Africa to French Polynesia, and Australia northwards to Japan.

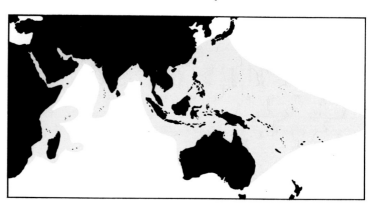

*Host anemone species*:

*Entacmaea quadricolor*

*Heteractis aurora*

*Heteractis crispa*

*Heteractis magnifica*

*Macrodactyla doreensis*

*Stichodactyla gigantea*

*Stichodactyla haddoni*

*Stichodactyla mertensii*

*D. trimaculatus* with *M. doreensis*, Mauritius. (G. Allen)

*D. trimaculatus*, adult, with *H. magnifica*, Bali, Indonesia. (G. Allen)

# Biology of Sea Anemones

## NUTRITION

Sea anemones that are host to clownfishes, like many tropical actinians and some temperate ones, harbour unicellular algae within the cells of their tentacles and oral disc (see Introduction). A portion of the sugars produced by these plants through photosynthesis are "leaked" to their host. This may be the anemone's major source of energy. The widely flared oral disc of many host actinians serves not only to accommodate fish, but its large surface area is well adapted for intercepting sunlight.

▲ *Stichodactyla mertensii* with *Amphiprion chrysopterus*, Solomon Islands. (R. Steene)

◄ *Heteractis magnifica* with *Amphiprion perideraion*, Flores, Indonesia. (G. Allen)

*Entacmaea quadricolor* can reproduce asexually, resulting in numerous individuals in close proximity. Clones of this anemone are frequently inhabited by *Amphiprion melanopus*. (H. Horn)

However, actinians, like all coelenterates, capture and digest animal prey with their nematocysts. We have found small fish, sea urchins, and a variety of crustaceans (shrimps and crabs) in the coelenteron of host anemones. They also appear to feed on planktonic items conveyed by the currents. Although the energy they derive from photosynthesis may be sufficient to live, the anemones need sulfur, nitrogen, and other elements in order to grow and reproduce. These animals are not voracious predators: their prey probably consists of animals that bump into them (e.g. a fish fleeing a more active predator) or stumble over them (e.g. a sea urchin, which has no eyes). Therefore, the supply is probably small and irregular. In hundreds of hours we have spent observing them we have never witnessed a host anemone feeding on fishes. A more predictable source of nutrients may be from wastes of their symbiotic fish. This issue deserves to be studied scientifically. Anemones of some species are capable of absorbing nutrients directly from seawater through their thin tissues, and that may be a source of nutrition for these animals as well.

## *SURVIVAL*

It is impossible to determine the age of a sea anemone, except for one that has been raised in an aquarium or tracked continuously in the wild from first settlement. A small one is not necessarily young, for coelenterates grow only if well fed and shrink if starved. Individuals of species that harbour clownfishes have been monitored for several years with no apparent change in size (although that is difficult to measure, due to the absence of a skeleton). However, studies on other species, in field and laboratory, have led to estimated ages on the order of many decades and even several centuries. There are scattered records of temperate anemones surviving many decades in commercial aquaria, and the life-span of a small sea anemone in New Zealand has been calculated, based on actuarial tables, to be over 300 years! From such data, it is likely that most individuals of the "gigantic" sea anemones we have encountered during our field work exceed a century in age. This is also consistent with the generalisation that large animals of all kinds typically are long-lived.

Coelenterates are well protected by their nematocysts, but some predators have developed means of evading their effect. Small tropical anemones may be eaten by butterflyfishes (see Chapter 5), but large ones appear to have few enemies. We do not know what might ultimately kill them.

## *REPRODUCTION*

All coelenterates reproduce sexually. An individual of some species may produce both eggs and sperm; host anemones appear to have separate sexes, with an individual being either male or female its entire life. The typical coelenterate pattern is that of most marine animals, one that is fraught with dangers and uncertainty — release of eggs and sperm into the sea, where fertilisation occurs and a larva (a tiny animal, looking

nothing like its parent, that drifts in the sea) develops for several days or weeks before settling in an appropriate habitat. Many species spawn in response to an environmental cue such as a full moon or low tide so that eggs and sperm are in the same place at the same time. Typically, marine animals produce millions of tiny larvae, but the world is not overrun with them, proving that very few survive — usually just enough to maintain a stable population. The rest of the larvae serve as food for a sea full of potential predators. Finally, the surviving larvae must find an appropriate habitat (how anemonefishes might do this is discussed in Chapter 4).

We do not know if host actinians follow this pattern. There is a bit of evidence that in at least some species, the eggs are not released, but are fertilised inside the mother (this is not especially rare in corals and anemones; sperm enter the mother with water that is constantly being pumped in and out, and which carries food and oxygen also), where they grow, to be released as tiny sea anemones. What is certain is that we seldom see small individuals of most host actinians in nature. However, it is not unusual to find large ones with ripe eggs and sperm. Therefore, we believe that successful recruitment must be rare. Very few eggs may be fertilised, or few larvae may survive, or larval settlement may be difficult, or young anemones may have high mortality (perhaps especially when they are too small to harbour fish). The apparent rarity of successful reproduction is also biologically consistent with long life.

In addition to sexual reproduction, some coelenterates undergo asexual reproduction. *Entacmaea quadricolor* is one of these. A polyp can divide longitudinally, resulting in two somewhat smaller individuals, probably within the space of a few days. Each then grows to an appropriate size, divides, and so on. All descendants of the original anemone (the result of sexual reproduction) form a clone, a group of genetically identical individuals. In this species, each polyp is relatively small, but clonemates remain next to one another so their tentacles are confluent, and the associated anemonefish apparently regard them as a single large anemone.

This is true mainly for shallow-water individuals; those in deeper water grow large, and do not divide (see Chapter 1). Several other species of actinians also have two different reproductive modes: small animals that clone and large ones that do not. This appears true of *Heteractis magnifica*, too. In the center of its range (i.e. in eastern Indonesia, on the Great Barrier Reef, and in New Guinea), it occurs as single, large individuals. To the east and west (i.e. in western Indonesia and Malaysia, and in Tahiti), several to very many small individuals of identical colouration are typically clustered together, appearing to be a single large (or huge!) anemone. Based on their shared colour and their proximity, we infer that they are clonemates.

## *LOCOMOTION*

Once they settle from the plankton, most anemones seldom move from place to place. Although they are usually damaged when people try to collect them, actinians do have the ability to detach from the substratum, partly or entirely. Small, temperate anemones can do this in response to predators or unfavorable physical factors. Indeed, those of a few species can "swim," awkwardly launching themselves into the water briefly, a motion that often puts them beyond reach of the predator that provoked the activity. More typically, an individual glides on its pedal disc, covering a few millimetres in a day, or it may detach entirely, and roll or be carried quite a distance. That this is not rare is attested by large animals suddenly appearing in well studied areas.

## ANEMONE-LIKE ANIMALS

Sea anemones are very similar to corals. One of the solitary mushroom corals, *Heliofungia actiniformis*, extends its tentacles by day (most do so only at night) and looks very much like an actinian (hence its specific name). Could it not harbour clownfishes? In an aquarium lacking an anemone, we did have an clownfish take up residence among a mushroom coral's tentacles. But it does not happen in nature. However, a small, snow-white pipefish, *Siokunichthys nigrolineatus*, lives among the tentacles, much like an anemonefish.

Corals and sea anemones differ not only in the respective presence and absence of a skeleton, but also in the types of nematocysts they possess and in their anatomy. Intermediate between them are corallimorpharians, skeletonless polyps having coral-like nematocysts and anatomy. In fact, some coelenterate experts regard corallimorpharians simply as corals without skeletons.

The corallimorpharian *Amplexidiscus fenestrafer*, the largest species known, resembles some host anemones. Bill Hamner first documented their feeding behaviour, in which prey is rapidly enveloped by the oral disc that closes like a draw-string purse. The mouth is then opened, the prey being swallowed and killed within the polyp. On several occasions he kept corallimorpharians in the same aquarium with anemonefishes, only to discover the following day that a fish was missing. He found that at night a fish settled onto the corallimorpharian's oral disc, just as it would with a host anemone, thereby provoking the draw-string response and subsequent ingestion. Their superficial similarity to some host anemones, and their living in areas of the reef where plankton accumulates, led us to speculate that these corallimorpharians might "lure" naive anemonefish larvae to attempt to settle in them. Of course, the possible mimicry of a host anemone, and the resultant predation on naive fry, requires that at least some anemonefish larvae recognize hosts at least in part visually (see Chapter 4).

The pipefish *Siokunichthys nigrolineatus* lives among the anemone-like tentacles of a mushroom coral, *Heliofungia actiniformis*. (G. Allen)

# III BIOLOGY OF SEA ANEMONES

*Amplexidiscus fenestrafer* is a corallimorpharian that greatly resembles a sea anemone. (H. Horn)

# Life History of Anemonefishes

The life history of many pomacentrids, and particularly *Amphiprion*, is well studied. Most research on anemonefishes has focused on *A. bicinctus*, *A. chrysopterus*, *A. clarkii*, *A. melanopus*, *A. ocellaris*, *A. perideraion*, and *A. tricinctus*, which are all similar in courtship and spawning, and in subsequent development of eggs, larvae, and young.

▲ Nests are located on solid objects near the anemone's column. A female *Amphiprion perideraion* lays her eggs as the male hovers nearby. (C. Roessler/Ikan)

# IV  LIFE HISTORY OF ANEMONEFISHES

A spawning pair of *Amphiprion clarkii*, Solomon Islands. (R. Steene)

## COURTSHIP, SPAWNING, AND INCUBATION

Within the tropics, spawning occurs throughout most of the year, although there may be seasonal peaks of activity. In subtropical or warm temperate seas (for example, in southern Japan), reproductive activity is generally restricted to spring and summer, when water temperatures are highest. At Enewetak Atoll (about $11^0$N in the central Pacific), spawning is strongly correlated with the lunar cycle: most nesting occurs when the moon is full or nearly so. Moonlight may serve to maintain a high level of alertness in the male, which assumes most of the nest guarding duties. Moreover, because newly hatched larvae are attracted to light, moonlight may draw them towards the surface, thereby facilitating their subsequent dispersal by waves and currents.

*Amphiprion* and *Premnas* are unique among damselfishes in forming permanent pair bonds that sometimes last for years. In other damsels, one male may mate with several females during a single spawning episode, and different sets of females are often involved in subsequent spawnings. However, pair-bonding in most species of clownfishes is very strong and is correlated with the small size of their territories (centred on actinians) which is, in turn, correlated with the unusual social hierarchy that exists in each "family" group. Details of this social structure will be given later.

Courtship in anemonefishes, as in all pomacentrids, is generally stereotyped and ritualised. Several days prior to spawning, there is increased social interaction, as expressed by chasing, fin-erection, and nest preparation. Another activity, which occurs in many damselfishes, is "signal jumping": the male swims rapidly up and down, as though on a roller-coaster. In anemonefishes, the male becomes particularly bold and

# LIFE HISTORY OF ANEMONEFISHES    IV

## *LIFE CYCLE OF AN ANEMONEFISH*

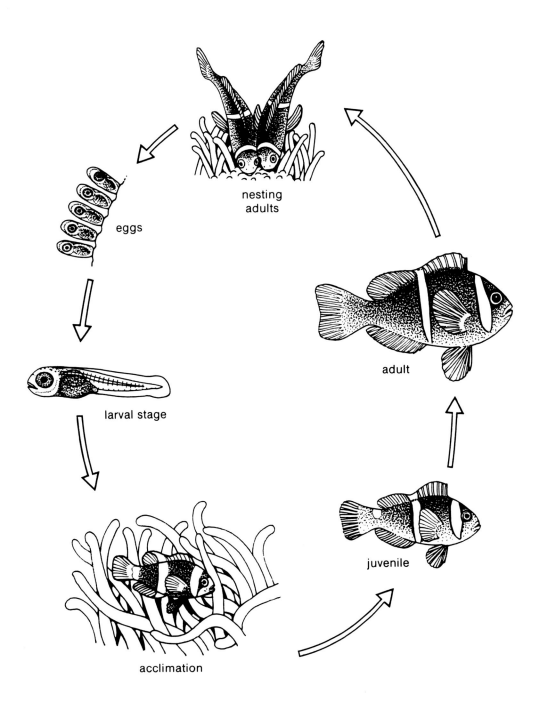

# IV  LIFE HISTORY OF ANEMONEFISHES

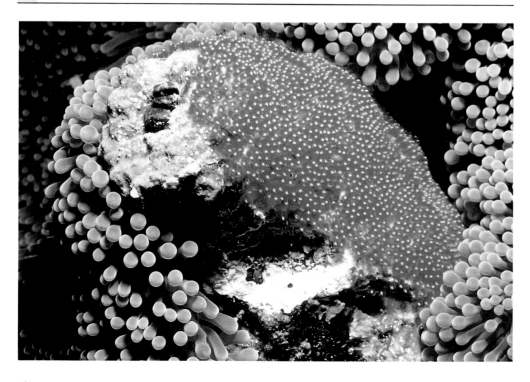

▲ Nest of *Amphiprion chrysopterus*. The anemone is *Stichodactyla mertensii*. (E. Robinson/Ikan)

▲ A pair of *Amphiprion chrysopterus* closely tend their nest of eggs. (E. Robinson/Ikan)

# LIFE HISTORY OF ANEMONEFISHES IV

A male *Amphiprion polymnus* fans its eggs. (G. Allen)

aggressive, chasing and nipping his mate. He also displays by fully extending his dorsal, anal, and pelvic fins, while remaining stationary in front of or beside her. During the nuptial period, he selects a nest site, usually on bare rock adjacent to the anemone. Initially the male spends considerable time clearing algae and debris from the site with his mouth; he is eventually joined in these activities by his mate.

Spawning, which occurs most often during morning hours, generally lasts from about 30 minutes to more than two hours. Once it commences, the tiny, conical ovipositor of the female is clearly visible. A number of eggs are extruded through this structure on each spawning pass, when the female swims slowly and deliberately in a zig-zag path with her belly just brushing the nest surface. She is followed closely by her mate, who fertilises the eggs as they are laid. Numerous passes occur during each spawning session. The number of eggs deposited ranges from about 100 to over 1000, depending on the size of the fish and on previous experience. In general, older, more experienced pairs produce more eggs than do recently formed pairs.

*Amphiprion* and *Premnas* eggs are elliptical or capsule-shaped, are about 3-4 mm in length, and adhere to the nest surface by a tuft of short filaments. They incubate six to seven days. Just prior to hatching, the embryo, which has undergone rapid development, is clearly visible through the transparent egg membrane: the most noticeable features are the large eyes with their silvery pupils, and the red-orange yolk sac that is responsible for the general colour of the entire egg mass when viewed from a short distance. Throughout incubation, the nest is meticulously guarded and cared for by the male. He aggressively chases other fishes from its vicinity, especially potential egg-eaters such as wrasses. He frequently visits the nest to fan the eggs with his pectoral fins and to remove dead eggs and debris with his mouth. The female is mainly occupied with feeding during this time, but occasionally assists the male with his duties.

## *LARVAL LIFE AND SETTLEMENT*

Hatching generally occurs during the evening, shortly after dark on the sixth or seventh day after the eggs are laid. In an aquarium, the freshly hatched fish first sink to the bottom, but within a few minutes swim to the upper part of the tank. The larvae are about 3-4 mm total length and transparent except for a few scattered pigmented spots, the eyes, and the yolk sac. Recent studies of larval duration in damselfishes have greatly improved knowledge of early life history stages. By counting the daily growth rings in the ear bones (otoliths) with an electron microscope, scientists can determine the time between hatching and transformation to the juvenile stage. There is much variation between species of damselfishes, with the longest larval stages about 6-8 weeks. Clownfishes have the shortest larval period of damsels, ranging from about 8 to 12 days. It is assumed that during this time they are planktonic — living in the surface waters of the ocean, where they are passively transported by currents. The short larval stage of anemonefishes is no doubt responsible for the localised distribution of many species.

The larval stage terminates when a young fish settles to the sea bottom and begins to assume the juvenile colour pattern. Aquarium observations indicate this metamorphosis is a rapid process, occurring within a day or so. At this stage it is vital for the young *Amphiprion* or *Premnas* to find a suitable anemone host or it will surely be consumed by one of its many predators. There is evidence that fish of some species can actively search and follow a trail of chemicals released by a host anemone, while others seem not to do this, and may locate a suitable host by sight, or simply encounter one by chance. For fish of some species, it takes several hours to become fully acclimated to the anemone once it is located; this is achieved by a series of progressively longer contacts with the tentacles like the elaborate "acclimation behaviour" seen when an adult fish is artificially removed from its host. Other fishes seem capable of swimming right in without harm, according to Miyagawa (see Chapter 5). Although she denied they go through "acclimation behaviour", she described swimming that resembles such behaviour. With 10 species of host actinians and 28 species of fish, there are probably many variations on how hosts are located and adapted to.

We assume that metamorphosis requires the presence of an anemone, since the fishes seem defenceless without one. We and others have done experiments proving that even adult anemonefishes cannot survive for long without the protection of a host actinian. What is obvious is that there are far fewer open slots available in appropriate anemones than there are fish to fill them. So there must be tremendous mortality among fry and larvae.

Even if it successfully locates an anemone, the immediate survival of the fish is not guaranteed. If the host is already occupied by anemonefish, the unusual social structure of the inhabitants makes life difficult for a newcomer. The number of fish that occupy a single anemone depends on

A host anemone is frequently occupied by an adult pair and several smaller individuals. *Amphiprion nigripes* and *Heteractis magnifica*, Maldive Islands. (Ikan)

species of fish, size of host, and sometimes size of the fish as well, but typically there is an adult pair and two to four smaller fish. As will be explained more fully below, the largest fish is usually the female and the next largest individual is her mate. A hierarchy, or "peck-order", exists in which the female is the dominant individual. There is generally an amicable relationship between the adult pair, and aggressiveness by the female is largely channeled into ritualised, non-harmful displays. Aggression is more overt farther down the hierarchy. The male spends considerable time chasing and "bullying" the next largest individual, which in turn vents its aggression on the smaller fish. Therefore, a new arrival becomes the immediate target for the resident fish. Attacks may be so severe as to drive away the newcomer, which must find another anemone or perish.

## SOCIAL STRUCTURE AND SEX REVERSAL

The phenomenon of sex reversal is an intriguing component of anemonefish life history, the details of which have been discovered only in the past two decades. Sex change occurs in many fishes. For example, it is now well established that most wrasses (Labridae) and parrotfishes (Scaridae) begin adult life as females and later assume the more colourful male phase. Similar changes are widespread among gropers (Serranidae), particularly in the subfamily Anthiinae, commonly known as fairy basslets. Therefore, it was not surprising to discover that this phenomenon extends to some pomacentrids as well. However, the unusual aspect of sex reversal in clownfishes is that the change is from male to female (protandrous hermaphroditism; the more common sort is protogynous hermaphroditism). As mentioned above, the largest and socially dominant fish in a particular anemone (or cluster of anemones in the case of *Entacmaea quadricolor*) is generally the female, whose gonads are functioning ovaries with remnants of degenerate testicular tissue. The smaller male, which in species such as *A. frenatus* and *P. biaculeatus* may be less than half the size of the female, has gonads that are functioning testes but also possess non-functioning or latent ovarian cells. If the dominant female dies or is experimentally removed, the male's gonads cease to function as testes and the egg-producing cells become active. Simultaneously, the largest of the non-breeding individuals becomes the functioning male. This adaptation allows continuous reproduction; without it, an adult would have to await the arrival of a fish of the appropriate sex (which it would be only 50% of the time), thereby losing valuable breeding time, or it would have to seek out a mate, leaving its anemone and thereby risking predation both on itself and on its symbiont.

There is no difference in colouration between sexes in most clownfishes, but there are exceptions, as shown on Page 51. The sex-related colour difference is not always present in *A. clarkii*. In addition, slight differences may occur between sexes of *A. perideraion* and *A. akallopisos*. Males of both species often have orange margins on the soft dorsal and caudal fins. In *P. biaculeatus*, colour differences may be related more to size than to sex *per se*.

## *FEEDING AND GROWTH*

Small, non-breeding fish are not necessarily young. The rigid social structure exerts a "stunting" effect on growth. Small fish use considerable energy fleeing from attacks of larger fish, and time that they might otherwise spend feeding is used in avoiding larger anemonefish. Moreover, it is especially unsafe for small individuals to range very far from the refuge of their anemone. Consequently, the small a fish, the less time it is able to devote to foraging, the more restricted its feeding area, and the more energy it must dedicate to evasionary activities. Members of the dominant pair range widely to feed, with fish of larger species, such as *A. clarkii*, swimming many metres from their actinian. When a fish is removed from the "queue", all those smaller than it grow rather rapidly, filling in the size gap. This is especially true of members of the mated pair: if the dominant female is removed, the new female (previously functioning as male) not only changes sex, but grows at an accelerated rate. The new male may grow even faster. Presumably this growth spurt is at least partly a result of a fish's being harassed less as it moves up the queue, and thereby having more time to feed.

Planktonic food is of major importance to most anemonefish. Copepods and larval tunicates are among the most common items found when their stomach contents are analysed. Fish of at least one species, *A. perideraion*, eat significant amounts of algae, which is both grazed from the surrounding reef and consumed in midwater.

Details of longevity are lacking for anemonefish of most species, but some are recorded to have lived at least 6-10 years in nature. The record for captive fish is 18 years, for *A. frenatus* and *A. perideraion* maintained at the Nancy Aquarium in France. The individual of *A. perideraion* was sill alive when this book was written.

This mixed pair consisting of a female *A. chrysopterus* and male *A. sandaracinos* are guarding a nest of eggs at Kimbe Bay, New Britain. (G. Allen)

## HYBRID ANEMONEFISHES

Two of the previously recognised species of *Amphiprion*, *A. leucokranos* and *A. thiellei*, are almost certainly hybrids. This conclusion is the result of field and aquarium observations over the past two years, after the first edition of this book appeared in 1992. Most of the evidence involves *A. leucokranos*, but is no doubt applicable to *A. thiellei*, which is virtually identical in colour and general appearance to reported variants of *A. leucokranos*. We are grateful to Dr. Bruce Carlson, Director of the Waikiki Aquarium in Honolulu, for sharing his information on this phenomenon. He was the first to suspect there was something unusual regarding the biology of *A. leucokranos*.

Unlike other anemonefishes, *A. leucokranos* is seldom found in pairs. Instead, individuals usually occur with those of other species, particularly *A. chrysopterus* and *A. sandaracinos*. In addition, there is far more colour variation among individuals of *A. leuckokranos* than in other clownfishes.

Recent surveys by both of us in Papua New Guinea and the Solomon Islands confirmed the hybrid theory. Relatively few individuals of *A. leukokranos* were observed, but they were invariably paired with either *A. chrysopterus* or *A. sandaracinos*. The most convincing evidence was the discovery of a mixed pair guarding a nest of eggs. The nest-guarding parents consisted of a large *A. chrysopterus* female and much smaller *A. sandaracinos* male. They were the only fish present in the anemone (*S. mertensii*). The eggs were clearly viable, and judging from their dark colouration were on the verge of hatching.

Pairs of "*leucokranos* hybrids" can successfully mate and produce offspring, but, judging from the relative scarcity of the fish in nature, it rarely occurs. Apparently they also back cross with the parent species, producing some very unusual offspring such as the fish shown on page 85 (bottom photo).

Breeding experiments currently underway at the Scripps Aquarium in San Diego, California, confirm that although *leucokanos* hybrids can successfully mate and produce viable young, the nest size is considerably smaller than normal, with only 50-100 eggs. So far there have been about a dozen spawnings, but only 1-10 fish have been successfully raised from each spawning. They exhibit tremendous colour variability, particularly regarding the white markings on the head. No two fish are alike - a few have the typical "bonnet" pattern with a disconnected white bar behind the eye (see page 87), others have the bar joined to the "bonnet", some have a small round spot instead of a bar behind the eye, still others have neither a spot or bar (only a "bonnet"), and some have a *sandaracinos*-like stripe on the forehead without any other marking.

# Interactions between Fish and Sea Anemones

This singular persistence of the fish to the same spot, and to the close vicinity of the great anemone, aroused in me strong suspicions of the existence of some connection between them.

These suspicions were subsequently verified... though what is the nature and object of that connection remains to be proved.

— Dr. Cuthbert Collingwood: *Rambles of a Naturalist on the Shores and Waters of the China Sea* (John Murray, London, 1868, pages 151-152)

More than a century later, people continue to be intrigued by that mysterious "connection." Two questions come immediately to mind. We

*Amphiprion nigripes* and *Heteractis magnifica*, Maldive Islands. (H. Debelius/Ikan)

# INTERACTIONS

Clownfishes are never found in nature without an anemone, *Amphirion leucokranos* with *Stichodactyla mertensii*, Solomon Islands. (B. and M. Jones/Ikan)

dealt in the Introduction with how small fish can survive in such a hostile environment. The other concerns the advantages and disadvantages of the relationship to the partners.

## *SYMBIOSIS*

The word symbiosis literally means "living together," implying no judgment about the benefits or detriments to either partner. In some countries, however, it is used as a synonym for "mutualism", a relationship from which both partners gain advantage. "Parasitism" is a type of symbiosis in which one partner benefits to the detriment of the other. There are two problems with invoking the latter terms. First, a relationship between organisms of two species may be either positive or negative, depending on circumstance. For example, a small amount of a virus can provoke immunity without disease, but a large amount can result in illness. The illness might be fatal in a very young or very old person. Thus, the "symbiosis" between a virus and a human may be good, indifferent, or harmful. Second, a great deal of understanding is necessary to assess the pluses and minuses accurately.

Recall that clownfishes are never found in nature without an anemone; this is an obligate association for them, although in captivity they are capable of living by themselves. It seems obvious that they are protected by the anemone with which they live — when threatened, they dive among its tentacles, from which most other fishes remain distant. We have taken clownfishes far from their anemones, and have removed anemones from beneath their fish. Poor swimmers, they sooner (in the former instance) or later (in the latter) became prey of larger fish. The presence of an anemone is also essential to reproduction of the fishes: their eggs are laid beneath the oral disc overhang of the anemone, where they are tended by the male (see Chapter 4).

# INTERACTIONS

In an aquarium, without an actinian, captive fish will bathe among air bubbles or frondose vegetation, so we infer that they obtain tactile stimulation from anemone tentacles. And the claim of some aquarists that the fish are livelier and healthier when kept with anemones suggests other benefits as well.

Indeed, aquarists have added much to knowledge of this symbiosis. Many have seen fish bring food to their anemones. This behaviour seems confined to aquaria. The normal diet of anemonefishes is small plants and animals that live in the water above the anemone, or algae that grow around it (Chapter 4). In nature, they do not encounter large particles of food, so they eat their food where it is found. Feeding large morsels to a fish in an aquarium produces an artifact: the fish, unable to devour the piece immediately, takes it home to work on it in the relative security of its own territory, as is typical of predators that obtain food in large amounts. But the territory in this case consumes the food!

Benefits or detriments to the anemone are not obvious. But neither do the fishes seem to harm their hosts. Therefore, many biologists have considered this a strictly one-sided relationship. The occasional anemones found in nature lacking fish support this conclusion — they seem to survive perfectly well without fish.

Or do they? On the Great Barrier Reef, at Enewetak Atoll, and in Papua New Guinea, we removed anemonefishes from hosts of the species *Entacmaea quadricolor* to determine how long it would take for new fish to repopulate them. But usually within 24 hours, there were no sea anemones at all! Instead, butterflyfish were poking their long snouts into crevices where the actinians had been anchored. In the absence of protection by their symbionts (which rarely happens in nature), anemones of this species fall prey to butterflyfishes. Hans Fricke made similar observations in the Red Sea, and Jack Moyer in Japan. Presumably, whatever protects butterflyfishes from the nematocysts of corals, on which they normally feed, does so as well for sea anemones. Thus, the relationship with its fishes is mutualistic for this species of anemone: the partners obviously profit by living together, protecting one another from predators.

Other, larger anemones can survive without fish, and apparently suffer no harm, but may do better when occupied by a clownfish. When threatened, a typical anemone withdraws its oral disc, closing its upper column over its tentacles. This is impossible in many host actinians because of their widely flared oral discs. Through evolutionary time, the anemone may have lost the ability to close because the fish protected it.

Other, minor benefits could be the fish eating parasites of the anemones, and fanning their hosts to create increased circulation of water over them. Clownfish do not, however, lure small fishes so that they can be eaten by the anemone, as was once believed. In fact, clownfish tend to keep other fish away from their anemones.

## SPECIFICITY

The relationship between fishes and anemones varies with the species of partner. One cannot speak of *the* anemone/fish symbiosis: there are nearly as many variants of it as there are combinations of species. To begin with, of the 28 fishes, only *Amphiprion clarkii* naturally occurs with all 10 host anemones; about a third of the fish are known from a single species of host, and the rest occur with several hosts (Appendix). Conversely, *E. quadricolor*, *H. crispa*, and possibly *S. mertensii* all host 12-14 fishes, whereas two play host to fish of only one species.

What governs which species occur together? Clearly, the fishes, being the shorter-lived and mobile partner, are responsible for this pattern. But how do they choose? Only species that occur in the same geographical area and have similar ecological preferences — sand or reef, deep or shallow — can potentially live together (although not all of them do). The sand-dwelling anemone *S. haddoni* provides an example of how ecology affects species specificity. *Amphiprion polymnus*, one of the largest and most aggressive of anemonefish, lives with it on relatively clean sand and in deep water. Where does such a fish lay its eggs? Egg clusters are attached to a solid object tucked beside the anemone's column. We have seen sand dollar tests, chunks of wood, and even a soft drink can or bottle used in this manner. In several instances, a groove in the clean sand extending some distance from the object on which the eggs had been laid provided evidence that the fish had dragged the object to the anemone. Because a hard substratum is necessary for reproduction, a fish large enough to drag an appropriate object to its anemone is required for success in that environment.

Beyond geographical and ecological coincidence, we believe three factors affect specificity between partners. 1) Fish have an innate or learned preference for only some of the anemones potentially available for colonisation. They may even choose based on habitat of the host: *A. melanopus* inhabits clones of *E. quadricolor* on tops of reefs, whereas *P. biaculeatus* lives in solitary specimens of *E. quadricolor* on reef slopes. 2) From among that subset of acceptable hosts, there is competition. The most host specific fish are those that are, generally, competitively superior for preferred anemones (for whatever reasons some anemones are preferred over others). 3) Chance.

If a fish's own mucus protects it from being stung, presumably it can occur only with actinians to which it has evolved a protective mucus. A fish larva settling into any anemone other than one of the "right" species would be killed. Indeed, some anemonefish are killed when placed in host actinians with which that fish species does not normally occur. There is evidence, however, that specificity is due to more than simply the deaths of all larvae that do not happen to settle into an appropriate host. Miyagawa found, in aquaria, that newly metamorphosed fry of some species locate an anemone by chemicals that are constantly being released by the anemone, much as a salmon senses its home stream, and

A nest of *Amphiprion polymnus* eggs on the base of a wine bottle. The anemone is *Stichodactyla haddoni*. (G. Allen)

that vision plays no role. These chemicals differ among species, so larval fish are attracted to anemones of species with which fish of that species naturally occur, but not to anemones of other species. However, fry of other fish are not attracted to anemones with which they naturally occur. So clownfishes may differ in how they select and locate hosts, as well as how they are protected from them.

If fish are protected from being stung by a coating of actinian mucus, it follows that they should be able to adapt to anemones with which they have had no previous contact (as an individual or a species). Indeed, a popular host anemone for home aquaria in the United States is the Caribbean species *Condylactis gigantea*, and some clownfishes can adapt to European or American actinians.

In either case, whether protection is from fish or anemone mucus, chemical attraction of larval fishes to host actinians is theoretically possible, and is consistent with some fishes being general in their host preferences and others being specific. The recognition of chemicals may be innate. Alternatively, it may be learned early in life. Recall, anemonefish eggs are incubated beside an actinian. During incubation, chemicals from the anemone may penetrate the egg case and imprint the embryonic fish. This is also analogous to how a salmon learns the "smell" of its home stream.

We believe that anemones having the greatest number of symbionts (10 or more in nature) are preferred by fishes for some as-yet-unknown reason. It may be that their chemical attractants are especially powerful, but that, in turn, may be a result of some other advantage they impart to their fish. Conversely, if it is advantageous to an anemone to have fish, they might have evolved a particularly potent attractant. In the equatorial tropics, *E. quadricolor* cannot live without fish; that anemone is among those with the greatest number of associates, which may assure it of having fish wherever it is. It is no coincidence that the two actinians associated with only a single species of fish both harbour *A. clarkii*, the least host-specific and most geographically widespread species of anemonefish. It owes its success, in terms of numbers and geographical extent, to its ability to occupy any host anemone.

Preference of fish for only certain anemones explains part of why there are only some species combinations in nature, but at least two other factors appear to have an influence. One is competition among fishes for anemones. Once fish of a particular species occupy an anemone, they chase out newcomers of other species, with rare exceptions. However, if an anemone is empty, and fish of different species settle in at about the same time, there can be competition. Fishes with greater host specificity are better competitors than those with less specificity. This is only reasonable: a fish that can live in only one species of anemone is entirely without refuge if it cannot prevent fish of other species from occupying that anemone. A fish with wide tolerances can be chased from one kind of anemone and still have others available.

Occasionally a fish known to be a poor competitor is found in a highly preferred anemone. We suspect the anemone missed being colonised by a member of the superior fish species, and so was available for colonisation by one of the inferior species. Once the resident fish grew a bit, it could prevent small individuals of other, perhaps competitively superior, species from occupying the anemone.

## *HOW ANEMONE AND FISH CAN AFFECT ONE ANOTHER*

Sea anemones of the species *Heteractis aurora* and *H. malu* seldom contain sexually mature pairs of fish; we term them "nursery anemones." Perhaps the hosts are inappropriate for large social groupings for some biochemical reason, although they are sufficiently hospitable that fish survival and growth to a certain size are possible. Given that a shortage of hosts limits natural anemonefish populations, we speculate that these actinians may allow resident fish the potential to move into actinians of more appropriate species when larger, jumping the queue, as it were. Some anemonefishes wander from their hosts, as juveniles or in southern Japanese waters when reproduction ceases in the winter. Fish are highly vulnerable to predation under such circumstances, but may survive sufficiently often for settlement into nursery anemones to be adaptive. Of course, a fish settling in such a host cannot know this, but (if it happens this way) the fish would have at least some chance of reproducing, whereas it would have none if it either remained in the "nursery anemone" its entire life, or, in the absence of space in a "better" species of host, settled into none at all. The basis for this effect of anemone upon fish is totally unknown.

In another instance of anemone affecting fish, the normally orange-coloured portions of the fish darken, so that the fish is black, rather than

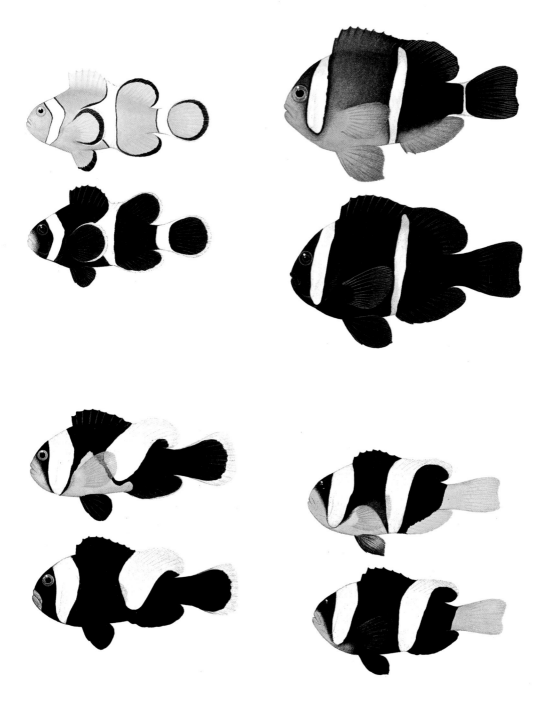

"Normal" and "melanistic" colour patterns for selected species of *Amphiprion*: Top left — *A. ocellaris*; top right — *A. trincinctus*; bottom left — *A. polymnus*; and bottom right — *A. sebae*.

# V INTERACTIONS

Melanistic variety of *Amphiprion clarkii* with *Stichodactyla mertensii*, Maldive Islands. (G. Allen)

orange, with white stripes. This type of melanism differs from that associated with size and certain isolated geographical populations (see Chapter 2). Only certain species of fish react this way, and only in certain species of actinians — for example, *A. chrysopterus* in *S. mertensii*, and *A. polymnus* in *H. crispa*. Such changes may be relatively rapid, so that an orange fish transferred to another anemone will darken within a matter of hours. Lightening, once removed from that host, generally occurs more slowly. The adaptive value of this reaction to either partner is unknown.

The fish can also affect its anemone. In the presence of a resident fish, tentacles of *E. quadricolor* bulge near the end, but in the absence of a fish, the tentacles commonly lack bulbs. Specimens of this anemone are often identified as different species based solely on tentacle form. But in all other respects they are indistinguishable. The transformation of an anemone from a member of the non-bulbous "species" into a member of the other putative species can be effected by placing an anemonefish among its tentacles, which develop bulbs within minutes. The reverse occurs when a fish is removed, although more slowly. The bulb exposes a larger surface area of the tentacle to sunlight, so that the algae may be able to gather more solar energy, but why that should happen only in the presence of fish, and how it occurs, are enigmas.

Several workers in the Red Sea have reported *A. bicinctus* diving inside *E. quadricolor*, although others have failed to find it. Some of the observations have been in the aquarium, where atypical behaviour often occurs. We have seen large adults of *A. polymnus* enter the mouth of *S. haddoni* under field conditions in Papua New Guinea. We do not know precisely what causes this behaviour — it does not always occur when the fish are pursued.

No doubt other interactions specific to particular combinations of anemones and fish will become clear as the symbiosis is investigated further, and names of both partners are used in a consistent manner.

# AQUARIUM CARE

CHAPTER SIX

# *Aquarium Care*

Anemonefishes have long ranked as one of the most popular attractions for marine aquarists. Bright colours, interesting behaviour, and the ability to adapt to captivity are mainly responsible for their popularity. However, relatively few species of anemonefishes enter the aquarium trade, largely owing to the remoteness and lack of commercial collectors in areas where many of them occur. The False Clown (*Amphiprion ocellaris*) is most typically seen in pet shops; huge quantities are regularly shipped from the Philippines, Indonesia, Singapore, and Hong Kong. Also common are Clark's, Tomato, and Spine-cheek Anemonefishes (*A. clarkii*, *A. frenatus*, and *Premnas biaculeatus*, respectively).

Fish of at least a few species, on the other hand, are now raised in large quantities on commercial fish farms. Aquarists who purchase only captive-bred fish need not be concerned about decimating natural populations. Indeed, breeding clownfishes offers a real challenge to the

▲ An aberrant colour form of *Amphiprion percula*, Solomon Islands. (R. Steene)

home aquarist. The main considerations for successful maintenance of anemonefish are provision of sufficient space, correct temperature, adequate filtration, proper water chemistry, ample shelter, and a nutritious diet. A number of excellent references available from pet stores, bookshops, and libraries adequately treat the subject of marine aquarium maintenance, which is therefore dealt with only briefly here.

## *AQUARIUM SIZE*

There are no hard and fast rules, despite published formulas (often based on a certain amount of water per inch or cm of fish). Common sense should prevail above all else, and generally, the largest possible aquarium should be used. Smaller species, such as *A. ocellaris* and *A. percula*, or juveniles obviously require less space than do adults of species such as *A. clarkii*. Although Clown Anemonefishes (*A. percula* and *A. ocellaris*) have successfully spawned in tanks as small as 50 litres, one of at least 300 litres should be provided for a pair of adult *A. clarkii* or similar-sized fish.

Clownfishes, like other damselfishes, are often aggressive in community situations. Anemonefish of about equal size, particularly those of a single species, frequently fight when kept together, as do adults; smaller specimens are generally compatible with other reef fishes. Aggressiveness is usually enhanced by the presence of an anemone, as the host is regarded as exclusive territory by every fish. Advanced aquarists who wish to breed their pets should set up pairs in isolation, or at least avoid community tanks with larger species.

## *TEMPERATURE*

Temperatures existing in their natural habitat are recommended for the well-being of clownfishes in captivity. Most species occur in the tropics, where the water generally ranges from about 25 to 28°C. Subtropical species such as *Amphiprion mccullochi* and *A. latezonatus* can be maintained in slightly cooler water (22-25°C).

## *FILTRATION AND WATER CHEMISTRY*

Filters on the market range from inexpensive subgravel types to sophisticated systems that monitor dissolved gas levels and cost more than $1,000. A few advanced aquarists, who usually live near the sea, utilise the so-called natural method, in which live corals and algae-covered rocks are maintained with little or no filtration. Regardless of filtration system used, a method of providing aeration and circulation (such as an airstone or bubbler) is essential. A minimal set-up for anemonefishes therefore must include a small electric air pump connected to a subgravel filter with a bubble tube and one or more airstones, depending on tank size. It is advisable to check nitrate levels and salinity occasionally, making necessary adjustments to achieve "normal" readings.

Beautiful blue variety of *Stichodactyla gigantea*, Madang, Papua New Guinea. Fish is *Amphiprion percula*. (R. Steene)

*Amphiprion clarkii* with rare purple variety of *Heteractis crispa*, Madang, Papua New Guinea. (G. Allen)

# AQUARIUM CARE

*Amphiprion perideraion* with violet form of *Heteractis magnifica*, Madang, Papua New Guinea. (G. Allen)

*Heteractis magnifica* and *Amphiprion perideraion*, Madang, Papua New Guinea. (G. Allen)

## DIET

In nature, anemonefish feed primarily on zooplankton (tiny animals, mainly crustaceans), supplemented with algae. This diet can easily be simulated, at least nutritionally, in captivity with finely chopped raw prawns and fish, and flake-type commercial fish food containing vegetable ingredients. This diet should be supplemented with live foods such as brine shrimp, *Daphnia*, mosquito larvae, and *Tubifex* worms whenever possible. However, freshwater organisms should be given in small quantities to prevent tank fouling. Several small feedings per day are better than a single large one.

## SHELTER

Shelter, in the form of rocks or small clay flower pots, is essential. If a rock shelter is provided, it is important to construct a small cave or crevice that can function as a retreat.

## ANEMONES

Captive fish can survive, and even breed, without anemones. Some clownfishes adapt successfully to actinians from places such as Hawaii, California, the Caribbean, and the Mediterranean, but, for some, this means keeping the aquarium cooler than is optimal for the fish and warmer than is normal for the anemone. The natural hosts are occasionally collected, and sporadically offered for sale, but this practice is to be discouraged. Natural hosts do not survive well (see below), necessitating frequent replacement. Local populations of *Amphiprion* and *Premnas* have become extinct on some reefs in the Philippines because of wanton destruction of habitat with dynamite (often to get fish) and removal of anemones by commercial collectors.

Elimination of the already limited number of suitable hosts not only deprives fishes of their homes, but it means that there is no source of more anemones! Actinians are long-lived in nature (see Chapter 3). Moreover, we have never seen very small individuals of most species, so successful reproduction is probably rare for them. It therefore appears that these lovely animals are being harvested much more rapidly than they are able to produce replacements.

Ignorance about their reproductive biology, combined with difficulty in maintaining them, preclude aquarium breeding of these actinians. In contrast to the fish, tropical sea anemones, including those with which clownfishes live, generally do not survive well under artificial conditions for reasons that are not entirely understood, but that probably have to do with nutrition. Most anemones lose their zooxanthellae after a short time in captivity. This phenomenon may be related to light, but the appropriate wavelengths and intensity are unknown, so the problem is difficult to remedy. It may also be related to the level of nitrogenous and other compounds in the water. Commercial aquaria that are successful at

# AQUARIUM CARE

Young of *Amphiprion percula* with *Heteractis magnifica*, Madang, Papua New Guinea. (G. Allen)

keeping corals (close kin of actinians that also possess zooxanthellae) circulate the water through beds of algae to remove such animal wastes, as happens in nature. This, of course, requires elaborate lighting and cooling. But even such enterprises generally end up with bleached anemones!

Lacking algae and probably a variety of natural nutrients, captive host anemones not only lose their normal colour, they assume shapes and postures not seen in nature. This can make them difficult to identify. For example, tentacles often shrink greatly, which probably means that the animal is living off its own tissues.

As mentioned previously, much useful information concerning the intricacies of the anemone and fish relationship has been gained through observation of captive specimens, particularly in large public aquaria. Even though we do not advocate the keeping of wild-caught tropical anemones, there is still much to be learned about the fishes. Amateur aquarists can certainly contribute to the growing fund of knowledge, particularly in the realm of clownfish reproduction, which has been adequately documented for relatively few species. Most serious aquarists maintain a logbook for recording notes on unusual and interesting behaviour. For those inclined to take this a step further there are a number of aquarium magazines that welcome contributions from the public. Even without these aspirations, the sheer enjoyment of keeping a tank of anemonefishes will no doubt ignite that same spark that Collingwood experienced on Fiery Cross Reef more than 120 years ago!

# APPENDIX

Appendix: Distribution of fishes among actinians; field records are lacking for *A. chagosensis*.

| Fish species* | Cryptodendrum Adhaesivum | Entacmaea quadricolor | Macrodactyla doreensis | Heteractis magnifica | Heteractis crispa | Heteractis aurora | Heteractis malu | Stichodactyla haddoni | Stichodactyla gigantea | Stichodactyla mertensii | Total number actinian associates |
|---|---|---|---|---|---|---|---|---|---|---|---|
| *Premnas biaculeatus* (Bloch) | | x | | | | | | | | | 1 |
| *Amphiprion ocellaris* Cuvier | | | | x | | | | | x | x | 3 |
| *Amphiprion percula* (Lacepède) | | | | x | | | | | x | x | 3 |
| *Amphiprion polymnus* (Linnaeus) | | | x | | x | | | x | | | 3 |
| *Amphiprion sebae* Bleeker | | | | | | | | x | | | 1 |
| *Amphiprion latezonatus* Waite | | | | | x | | | | | | 1 |
| *Amphiprion akallopisos* Bleeker | | | | x | | | | | | x | 2 |
| *Amphiprion nigripes* Regan | | | | x | | | | | | | 1 |
| *Amphiprion perideraion* Bleeker | | | x | x | | | | | x | | 4 |
| *Amphiprion sandaracinos* Allen | | | | x | x | | | | | x | 2 |
| *Amphiprion leucokranos* Allen | | | | x | x | | | | | x | 3 |
| *Amphiprion ephippium* (Bloch) | | x | | | x | | | | | | 2 |
| *Amphiprion frenatus* Brevoort | | x | | | | | | | | | 1 |
| *Amphiprion mccullochi* Whitley | | x | | | | | | | | | 1 |
| *Amphiprion melanopus* Bleeker | | x | | x | x | | | | | | 3 |
| *Amphiprion rubrocinctus* Richardson | | x | | | | | | | x | | 2 |
| *Amphiprion clarkii* (Bennett) | x | x | x | x | x | x | x | x | x | x | 10 |
| *Amphiprion akindynos* Allen | | x | | | | x | | x | x | x | 6 |
| *Amphiprion allardi* Klausewitz | | x | | | | x | | | | x | 3 |
| *Amphiprion bicinctus* Rüppell | | x | | x | x | x | | | x | x | 6 |
| *Amphiprion chagosensis* Allen | | | | | | | | | | | ? |
| *Amphiprion chrysogaster* Cuvier | | | x | x | | x | | x | | x | 5 |
| *Amphiprion chrysopterus* Cuvier | | x | | x | x | x | | x | | x | 6 |
| *Amphiprion fuscocaudatus* Allen | | | | | | | | | | x | 1 |
| *Amphiprion latifasciatus* Allen | | | | | | | | x | | | 1 |
| *Amphiprion omanensis* Allen and Mee | | x | | | x | | | | | | 3 |
| *Amphiprion tricinctus* Schultz and Welander | | x | | | x | x | | | | x | 3 |
| Total number obligate fish associates | 1 | 13 | 4 | 11 | 14 | 7 | 1 | 7 | 7 | 13 | |
| *Dascyllus trimaculatus* Rüppell | | x | x | x | x | x | x | x | x | x | 8 |
| *Dascyllus albisella* Gill | | | | | | | | | | | |

* Genera and species complexes of *Amphiprion* (after Allen, 1972) separated by horizontal lines.

# ACKNOWLEDGEMENTS

This is contribution number 66 of the Christensen Research Institute, Madang, Papua New Guinea. A generous grant from the Christensen Fund gave us the impetus to begin this project, which we had been discussing for a decade. We also acknowledge with gratitude the Director and Board of Trustees of the Western Australian Museum for making the publication of this book possible.

Fellowships to both of us from the CRI contributed to this work. Grants to DGF from the U. S. National Science Foundation (DEB 76-82277), the U. S. National Academy of Sciences (Bache Fund grant 551), the National Geographic Society (grant 1741), the Cocos Foundation, and the California Academy of Sciences are gratefully acknowledged.

Individuals too numerous to mention provided assistance to DGF in the field, museum, and laboratory. She is especially indebted to Jack T. Moyer (Tatsuo Tanaka Memorial Biological Station, Miyake-jima, Japan); Aprilani Soegiarto (formerly of Lembaga Oseanologi Nasional, Jakarta, Indonesia) and his staff including M. Hutomo, Kassim Moosa, and the crew of the R.V. *Samudera*; Richard N. Mariscal (Florida State University); and John Mizeu and Jean Pierret (CRI). Professor B. Condé, Director of the Nancy Aquarium (Musée de Zoologie de l'Université de Nancy) kindly provided longevity data for captive *Amphiprion*.

Talented Perth natural history artist Roger Swainston is responsible for the exquisite paintings that illustrate this work. We were also fortunate to obtain a selection of excellent slides from the Ikan Agency in Frankfurt, Germany, under the auspices of Helmut Debelius. Individual photographers are given credit. Special mention should be given Roger Steene, one of Australia's foremost underwater photographers, for providing a number of photos and assisting GRA on numerous field trips. Valuable photographs were also obtained from R. Eisenhart, H. Horn, and L. Preston. Finally, we thank Jill Ruse for preparing the line drawings.

We thank the capable staff of the Publications Department of the Western Australian Museum who greatly assisted us during the production stages: Greg Jackson, Ann Ousey, Desmond Doherty, Vince McInerney and Malcolm Parker.

# REFERENCES AND RECOMMENDED READING

Allen, G. R. 1972. *The Anemonefishes: their Classification and Biology.* T. F. H. Publications, Inc., Neptune City, New Jersey, 288 pages.
Allen, G. R. 1973. *Amphiprion leucokranos,* a new species of pomacentrid fish, with notes on other anemonefishes of New Guinea. Pacific Science 274: 319-326.
Allen, G. R. 1975. Anemonefishes and their amazing partnership. Australian Natural History 18(8): 274-277.
Allen, G. R. 1978. *Die Anemonen-fische Arten der Welt.* Mergus Verlag, Melle, Germany, 104 pages.
Allen, G. R. 1980. *Anemonefishes of the World: Species, Care, and Breeding.* Aquarium Systems, Mentor, Ohio, 104 pages.
Allen, G.R. 1991. *Damselfishes of the World.* Mergus Verlag, Melle, Germany, 271 pages.
Allen, G. R. and R. N. Mariscal. 1971. A redescription of *Amphiprion nigripes* Regan, a valid species of anemonefish (family Pomacentridae) from the Indian Ocean. Fieldiana, Zoology 58(8): 93-101.
Bell, L. J. 1976. Notes on the nesting success and fecundity of the anemonefish *Amphiprion clarkii* at Miyake-Jima, Japan. Japanese Journal of Ichthyology 22(4): 207-211.
Bell, L. J., J. T. Moyer, and K. Numachi. 1982. Morphological and genetic variation in Japanese populations of the anemonefish *Amphiprion clarkii.* Marine Biology 72: 99-108.
Brooks, W. R. and R. N. Mariscal. 1984. The acclimation of anemone fishes to sea anemones: protection by changes in the fish's mucous coat. Journal of Experimental Marine Biology and Ecology 81: 277-285.
Collingwood, C. 1868. Note on the existence of gigantic sea-anemones in the China Sea, containing within them quasi-parasitic fish. Annals and Magazine of Natural History, series 4, 1: 31-33.
Crespigny, C. C. de. 1869. Notes on the friendship existing between the malacopterygian fish *Premnas biaculeatus* and the *Actinia crassicornis.* Proceedings of the Zoological Society of London 1869: 248-249.
Cutress, C. E. 1977. Orders Corallimorpharia, Actiniaria, and Ceriantharia. In: *Reef and Shore Fauna of Hawaii, Section 1: Protozoa through Ctenophora.* D. M. Devaney and L. G. Eldredge, editors. Bernice P. Bishop Museum Special Publication 64, Honolulu, Hawaii, pages 130-147.
Cutress, C. E. and A. C. Arneson. 1987. Sea anemones of Enewetak Atoll. In: *The Natural History of Enewetak Atoll.* D. M. Devaney, E. S. Reese, B. L. Burch, and P. Helfrich, editors. Office of Science and Technology Information, U. S. Department of Energy, Washington, D.C., pages 53-61 (chapter 6).

# REFERENCES AND RECOMMENDED READING

Davenport, D. and K. S. Norris. 1958. Observations on the symbiosis of the sea anemone *Stoichactis* and the pomacentrid fish, *Amphiprion percula*. Biological Bulletin 115: 397-410.

Doumenc, D. 1973. Notes sur les actinies de Polynêsie Française. Cahiers du Pacifique 17: 173-204.

Dunn, D. F. 1974. *Radianthus papillosa* (Coelenterata, Actiniaria) redescribed from Hawaii. Pacific Science 28(2): 171-179.

Dunn, D. F. 1981. The clownfish sea anemones: Stichodactylidae (Coelenterata: Actiniaria) and other sea anemones symbiotic with pomacentrid fishes. Transactions of the American Philosophical Society 71: 1-115.

Fautin, D. G. 1985. Competition by anemone fishes for host actinians. Proceedings of the Fifth International Coral Reef Congress, Tahiti 5: 373-377.

Fautin, D. G. 1986. Why do anemonefishes inhabit only some host actinians? Environmental Biology of Fishes 15(3): 171-180.

Fautin, D. G. 1987. Who are those little orange fish and why are they living in a sea anemone? Pacific Discovery 40(2): 18-29.

Fautin, D. G. 1988. Sea anemones of Madang Province. Science in New Guinea 14(l): 22-29.

Fautin, D. G. 1989. Sexual stunts of clownfish. Natural History 9/89: 42-47.

Fautin, D. G. 1991. The anemonefish symbiosis: what is known and what is not. Symbiosis 10: 23-46.

Fautin, D. C. 1992. Anemonefish recruitment: the roles of order and chance. Symbiosis 14: 143-160.

Fishelson, L. 1965. Observations and experiments on the Red Sea anemones and their symbiotic fish *Amphiprion bicinctus*. Bulletin of the Israel Sea Fisheries Research Station, Haifa 39: 3-14.

Fishelson, L. 1970. Protogynous sex reversal in the *Anthias squamipinnis* (Teleostei, Anthiidae) regulated by presence or absence of male fish. Nature (London) 227: 90-91.

Fricke, H. W. 1973. Individual partner recognition in fish: field studies on *Amphiprion bicinctus*. Naturwissenschaften 60: 204-205.

Fricke, H. W. 1974. Oko-ethologie des monogamen Anemonenfisches *Amphiprion bicinctus* (Freiwasseruntersuchung aus dem Roten Meer). Zeitschrift für Tierpsychologie 36: 429-412.

Fricke, H. W. 1975. Selektives Feinderkennen bei dem Anemonenfisch *Amphiprion bicinctus* (Rüppell). Journal of Experimental Marine Biology and Ecology 19: 1-7.

Fricke, H. W. 1975. Sozialstruktur und ökologische Spezialisierung von verwandten Fischen (Pomacentridae). Zeitschrift für Tierpsychologie 39: 492-520.

Fricke, H. W. 1977. Community structure, social organization and ecological requirements of coral reef fish (Pomacentridae). Helgoländer wissenschaftlichen Meeresuntersuchungen 30: 412-426.

Fricke, H. W. 1979. Mating system, resource defence and sex change in the anemonefish *Amphiprion akallopisos*. Zeitschrift für Tierpsychologie 50: 313-326.

Fricke, H. W. 1983. Social control of sex: field experiments with the anemonefish *Amphiprion bicinctus*. Zeitschrift für Tierpsychologie 61: 71-77.

Fricke, H. and S. Fricke. 1977. Monogamy and sex change by aggressive dominance in coral reef fish. Nature 266: 830-832.

Godwin, J. and D. G. Fautin. 1992. Defense of host actinians by anemonefishes. Copeia, 3: 902-907.

Gohar, H. A. F. 1948. Commensalism between fish and anemone (with a description of the eggs *of Amphiprion bicinctus* Rüppell). Publications of the Marine Biological Station, Ghardaqa, Egypt 6: 35-44.

# REFERENCES AND RECOMMENDED READING

Hanlon, R. T. and R. F. Hixon. 1986. Behavioral associations of coral reef fishes with the sea anemone *Condylactis gigantea* in the Dry Tortugas, Florida. Bulletin of Marine Science 39(1): 130-134.

Lubbock, R. 1980. Death where is thy sting? New Scientist 4: 153-154.

Lubbock, R. 1980. Why are clownfishes not stung by sea anemones? Proceedings of the Royal Society of London 207: 35-61.

Mariscal, R. N. 1969. An experimental analysis of the protection of *Amphiprion xanthurus* Cuvier & Valenciennes and some other anemone fishes from sea anemones. Journal of Experimental Marine Biology and Ecology 4: 134-149.

Mariscal, R. N. 1970. A field and laboratory study of the symbiotic behavior of fishes and sea anemones from the tropical Indo-Pacific. University of California Publications in Zoology 91: 1-43.

Mariscal, R. N. 1970. The nature of the symbiosis between Indo-Pacific anemone fishes and sea anemones. Marine Biology 6: 58-65.

Mariscal, R. N. 1972. Behavior of symbiotic fishes and sea anemones. In: *Behavior of Marine Animals*, volume 2. H. E. Winn and B. L. Olla, editors. Plenum Publishing Corporation, New York, pages 327-360 (chapter 9).

Masry, D. 1971. A photographic note on the intimate relationship between a young symbiotic fish *Amphiprion bicinctus* Ruppell [sic] and its anemone host at Elat (Red Sea). Israel Journal of Zoology 20: 139-142.

Miyagawa, K. and T. Hidaka. 1980. *Amphiprion clarkii* juvenile: innate protection against and chemical attraction by symbiotic sea anemones. Proceedings of the Japan Academy 56B(6): 356-361.

Miyagawa, K. 1989. Experimental analysis of the symbiosis between anemonefish and sea anemones. Ethology 80: 19-46.

Moyer, J. T. 1976. Geographical variation and social dominance in Japanese populations of the anemonefish *Amphiprion clarkii*. Japanese Journal of Ichthyology 23(1): 12-22.

Moyer, J. T. 1980. Influence of temperate waters on the behavior of the tropical anemonefish *Amphiprion clarkii* at Miyake-Jima, Japan. Bulletin of Marine Science 30: 261-272.

Moyer, J. T. and L. J. Bell. 1976. Reproductive behavior of the anemonefish *Amphiprion clarkii* at Miyake-Jima, Japan. Japanese Journal of Ichthyology 23(1): 23-32.

Moyer, J. T. and A. Nakazono. 1978. Protandrous hermaphroditism in six species of the anemonefish genus *Amphiprion* in Japan. Japanese Journal of Ichthyology 25(2): 101-106.

Moyer, J. T. and C. E. Sawyers. 1973. Territorial behavior of the anemonefish *Amphiprion xanthurus* with notes on the life history. Japanese Journal of Ichthyology 20(2): 85-93.

Moyer, J. T. and R. C. Steene. 1979. Nesting behavior of the anemonefish *Amphiprion polymnus*. Japanese Journal of Ichthyology 26(2): 209-214.

Murata, M., K. Miyagawa-Kohshima, K. Nakanishi, and Y. Naya. 1986. Characterization of compounds that induce symbiosis between sea anemone and anemone fish. Science 234: 585-587.

Ochi, H. 1986. Growth of the anemonefish *Amphiprion clarkii* in temperate waters, with special reference to the influence of settling time on the growth of 0-year olds. Marine Biology 92: 223-229.

Reed, S. A. 1971. Some common coelenterates in Kaneohe Bay, Oahu, Hawaii. In: *Experimental Coelenterate Biology*. H. M. Lenhoff, L. Muscatine, and L. V. Davis, editors. University of Hawaii Press, Honolulu, pages 37-51.

Ross, R. M. 1978. Reproductive behavior of the anemonefish *Amphiprion melanopus* on Guam. Copeia 1978: 103-107.

# REFERENCES AND RECOMMENDED READING

Ross, R. M. 1978. Territorial behavior and ecology of the anemonefish *Amphiprion melanopus* on Guam. Zeitschrift für Tierpsychologie 46: 71-83.

Saville-Kent, W. 1893. *The Great Barrier Reef of Australia: Its Products and Potentialities*. W. H. Allen and Co., London. 387 pages.

Schlichter, D. 1968. Der Nesselschutz der Anemonenfische. Verhandlungen der Deutschen Zoologischen Gesellschaft in Innsbruck 32: 327-333.

Schlichter, D. 1968. Das Zusammenleben von Riffanemonen und Anemonenfischen. Zeitschrift für Tierpsychologie 25: 933-954.

Schlichter, D. 1972. Chemische Tarnung. Die stoffliche Grundlage der Anpassung von Anemonenfischen und Riffanemonen. Marine Biology 12: 137-150.

Schlichter, D. 1976. Macromolecular mimicry: substances released by sea anemones and their role in the protection of anemone fishes. In: *Coelenterate Ecology and Behavior*. G.O. Mackie, editor. Plenum Press, New York and London, pages 433-441.

Uchida, H., K. Okamoto, and T. Fukuda. 1975. Some observations on the symbiosis between anemonefishes and sea-anemones in Japan. Bulletin of the Marine Park Research Stations 1(1): 31-46.

Verwey, J. 1930. Coral reef studies. I. The symbiosis between damselfishes and sea anemones in Batavia Bay. Treubia 12: 305-366.

Wood, E. M. Behaviour and social organization in anemonefish. Progress in Underwater Science 11: 53-60.

# *GLOSSARY*

actinian — sea anemone, from the scientific term for sea anemones, Actiniaria

*Amphiprion* — fish genus containing 27 species, all of which are symbiotic with sea anemones

cnidarian — member of phylum Cnidaria (also called Coelenterata), the group to which sea anemones, corals, jellyfish, and hydras belong

coelenterate — member of phylum Coelenterata (also called Cnidaria); see cnidarian

*Cryptodendrum* — sea anemone genus containing only one species, which may host anemonefish

*Dascyllus* — genus of damselfishes, some species of which are facultative symbionts of actinians, particularly when young

*Entacmaea* — genus containing at least two species, one of which is *E. quadricolor*, the most abundant host sea anemone, which harbours the greatest number of species of clownfishes

facultative symbiosis — a relationship in which one partner may, but need not, live with another in order to survive; the relationship of most species of host sea anemones with anemonefishes is facultative

*Heteractis* — genus containing four species of sea anemones, all of which may host anemonefish

host — one of the partners in a symbiosis, generally the larger one (thus, the sea anemones that are the subject of this book are hosts to fish, crustacean, and algal symbionts)

larva (plural larvae) — a developmental stage that hatches from an egg and that typically lives in a different environment, looks entirely different, and eats different food from the adult of the species; oceanic larvae are generally very small, and are planktonic

*Macrodactyla* — sea anemone genus containing only one species, which may host anemonefish

metamorphosis — the change from larva to adult

melanism — a tendency to develop black pigmentation, common in many species of anemonefish

mutualism — a symbiosis in which both partners benefit

nematocyst — microscopic, harpoon-like stinging capsule manufactured by all cnidarians (= coelenterates)

obligate symbiosis — a relationship in which one partner *must* live with another in order to survive; except for a brief planktonic larval stage, anemonefishes are obligate symbionts of sea anemones

# GLOSSARY

parasitism — a symbiosis in which one partner benefits, to the detriment of the other

plankton — oceanic plants and animals that drift with the currents; although most planktonic species are small, some, such as jellyfishes, are not

*Premnas* — genus containing only one species, which is an anemonefish

*Stichodactyla* — genus containing five species of sea anemones, three of which may host anemonefish

symbiont — one of the partners in a symbiosis

symbiosis — literally "living together", used by scientists to describe the relationship between unrelated species of plants and/or animals that live in intimate association

taxonomy — the naming of living beings based on their evolutionary relationships to one another

zoophyte — literally "animal plant", an archaic term for a sea anemone

zooxanthellae — single-cell, dinoflagellate (golden-brown) algae that live symbiotically within the cells of some marine animals such as most reef-forming corals, many tropical and a few temperate sea anemones, some hydroids, and all giant clams

# INDEX

Page numbers appearing in bold print refer to the primary species account of the fish or anemone. Italics refer to illustrations.

actiniformis, *Heliofungia* 123
*Actinia*
  *aurora* 28
  *crispa* 30
  *doreensis* 36
  *magnifica* 32
  *quadricolor* 26
actinostoloides,
  *Cymbactis* 26
  *Parasicyonis* 26
actinostoroides, *Parasicyonis* 26
adhaesivum, *Cryptodendrum* 8, *11*, 19, 22, **24**, *25*, 74
Adhesive Sea Anemone 24
akallopisos, *Amphiprion* 33, 42, 46, 53, **60**, *61*, *100*, 106, 132
akindynos, *Amphiprion* 26, 29, 30, 39, 41, 42, 43, 48, 56, **62**, *63*, 68, 72, 74
albisella, *Dascyllus* 45, 116
allardi, *Amphiprion* 26, 29, 43, 48, 57, **64**, 65, 66, 68, 74, 84
Allard's Anemonefish 64
amblycephalus, *Thalassoma* 5
*Amphiprion*
  akallopisos 33, 42, 46, 53, **60**, *61*, *100*, 106, 132
  akindynos 26, 29, 30, 39, 41, 42, *43*, 48, 56, **62**, *63*, 68, 72, 74
  allardi 26, 29, 43, 48, 57, **64**, *65*, 66, 68, 74, 84
  bicinctus *1*, 27, 29, 30, 33, 39, *47*, 57, **66**, *67*, 68, 84, 125, 142
  chagosensis 9, 48, 56, 66, **68**, *69*, 74
  chrysogaster 29, 36, 41, 43, 47, 48, 59, **70**, *71*, 74, 80, 112
  chrysopterus 23, 27, 29, 31, 33, 41, 43, 48, 49, 56, 57, 62, 64, **72**, *73*, *119*, *125*, *128*, 133, 142
  clarkii 2, 6, 9, 11, 24, 27, 29, 31, 33, 34, 35, 36, 37, 39, 41, 43, 46, 49, 50, 51, 55, 57, 62, 72, **74**, *75*, 84, 125, *126*, 132, 138, 142, 143, 144, *145*
  ephippium 27, 31, *47*, 48, 52, **76**, *77*
  frenatus 27, 48, 51, 54, 76, **78**, *79*, 90, 104, 131, 143
  fuscocaudatus 43, 59, 70, 74, **80**, *81*, 112
  latezonatus 31, 46, 59, **82**, *83*, 144
  latifasciatus 43, 56, 64, 66, 74, **84**, *85*, 96
  leucokranos 31, 33, 46, 49, 54, **86**, *87*, *100*, 106, 133, *136*
  mccullochi 27, 52, 53, **88**, *89*, 144
  melanopus 27, 31, 33, 46, *47*, 48, 49, 52, 53, 54, 76, 78, 88, **90**, *91*, 120, 121, 125, 138
  nigripes 33, *46*, 52, 53, **92**, *93*, *100*, *110*, *131*, *135*
  ocellaris 33, 39, 43, 46, 58, **94**, *95*, 98, 125, *141*, 143, 144
  omanensis 27, 31, 56, 84, **96**, *97*
  percula *10*, 31, 33, *39*, *45*, 46, 49, 58, 94, **98**, *99*, *143*, 144, *145*, 148
  perideraion 2, *13*, *17*, *31*, *33*, 36, 39, 44, 46, 54, 92, 100, 101, *110*, *125*, *132*, *146*
  polymnus 8, *31*, 41, 48, 55, **102**, 103, *108*, *129*, *138*, *139*, *141*, *142*
  rubrocinctus 27, 39, 48, 54, 76, 78, 90, 104, *105*
  sandaracinos 43, 46, 53, 60, 86, 100, **106**, *107*, *133*
  sebae 41, 48, 55, 102, 108, *109*, *141*
  thiellei 54, 110, 111, 133
  tricinctus 27, 29, 31, 43, 48, 55, 59, 70, 72, 74, **112**, 113, *125*, *141*
Amphiprioninae 14, 45
*Amplexidiscus fenestrafer* 123, 124
Anemonefish
  Allard's 64
  Australian 104
  Barrier Reef 62
  Chagos 68
  Clark's 74
  Clown 98
  False Clown 94
  Madagascar 84
  Maldives 92
  Mauritian 70
  McCulloch's 88
  Oman 96
  Orange 106
  Orange-fin 72
  Pink 100
  Red and Black 90
  Red Saddleback 76
  Saddleback 102
  Sebae 108
  Seychelles 80
  Skunk 60
  Spine-cheek 114
  Thielle's 110
  Three-band 112
  Tomato 78
  Two-band 66
  White-bonnet 86
  Wide-band 82
*Antheopsis papillosa* 34
*Anthias clarkii* 74
aurora,
  *Actinia* 28

# INDEX

*Heteractis* 19, 22, **28**, *29*, 30, 34, 36, 62, 64, 66, 70, 72, 74, 112, 116
Australian Anemonefish 104
Barrier Reef Anemonefish 62
*Bartholomea* sp. 28
Beaded Sea Anemone 28
*biaculeatus*,
    *Chaetodon* 114
    *Premnas* 6, 11, 27, 48, 51, 59, 114, 115, 131, 132, 138, 143
*bicinctus, Amphiprion* 1, 27, 29, 30, 33, 39, *47*, 57, **66**, *67*, 68, 84, 125, 142
Bulb-Tentacle Sea Anemone 26
*Chaetodon biaculeatus* 114
Chagos Anemonefish 68
*chagosensis, Amphiprion* 9, 48, 56, 66, **68**, *69*, 74
Chrorninae 45
*chrysogaster, Amphiprion* 29, 36, 41, 43, 47, 48, 59, **70**, 71, 74, 80, 112
*chrysopterus, Amphiprion* 23, 27, 29, 31, 33, 41, 43, 48, 49, 56, 57, 62, 64, **72**, *73*, *119*, *125*, *128*, *142*
*clarkii, Amphiprion* 2, 6, 9, 11, 24, 27, 29, 31, 33, 34, 35, 36, 37, 39, 41, 43, 46, 49, 50, 51, 55, 57, 62, 72, **74**, *75*, 84, 125, *126*, 132, 138, 142, 143, 144, *145*
Clark's Anemonefish 74
Clown Anemonefish 98
*Condylactis gigantea 139*
*cookei, Macranthea* 34
Corkscrew-Tentacle Sea Anemone 36
*crispa*,
    *Actinia* 30
    *Heteractis* 1, 18, 20, 21, 22, 28, **30**, 31, 34, 36, 62, 63, 66, 72, 74, 75, 76, *77*, *82*, *83*, 86, 87, 90, 96, 98, 100, 102, 112, 116, 138, 142, *145*
*Cryptodendrum adhaesivum* 8, *11*, *19*, *22*, 24, 25, 74
*Cymbactis actinostoloides* 26
Dascyllas, Three-Spot 116
*Dascyllas trimaculatus* 8, *37*, 45, 48, **116**, *117*
Delicate Sea Anemone 34
*digitata, Stoichactis* 24,
*Discosoma*
    *giganteum* 38
    *haddoni* 40
    *malu* 34
*doreensis*,
    *Actinia* 36
    *Macrodactyla* 6, 20, 22, 28, **34**, 36, 37, 70, 74, 100, 116, *117*
*douglasi, Physobrachia* 26
*Entacmaea quadricolor* 11, 18, 19, 21, 26, 27, 62, *63*, 64, 65, 66, 74, 76, 78, *79*, *88*, *89*, *90*, *91*, *96*, *97*, 104, *105*, *112*, *113*, *114*, *115*, *116*, *121*, 122, 131, 137, 138, 139, 142
*ephippium*,
    *Amphiprion* 27, 31, *47*, 48, 52, **76**, *77*
    *Lutjanus* 76
False Clown Anemonefish 94
*fenestrafer, Amplexidiscus* 123, *124*
*frenatus, Amphiprion* 27, 48, 51, 54, 76, **78**, *79*, *90*, 104, 131, 143

*fuscocaudatus, Amphiprion* 43, 59, 70, 74, 80, 81, 112
*gelam*,
    *Macrodactyla* 36
    *Radianthus* 26, 36
*gigantea*,
    *Stichodactyla* **9**, *10*, *20*, **22**, 38, 39, 40, 42, 66, 74, 94, 98, *99*, *100*, 104, 116, *145*
    *Stoichactis* 40, 42
*giganteum, Discosoma* 38
*giganteus, Priapus* 38
Gigantic Sea Anemone 38
*Gyrostoma*
    *helianthus* 26
    *quadricolor* 26
Haddon's Sea Anemone 40
*haddoni*,
    *Discosoma* 40
    *Stichodactyla* 8, 9, 10, 32, 38, **40**, 41, 42, 62, 70, 72, 74, 102, *103*, 108, 116, 138, *139*, 142
    *Stoichactis* 40
*helianthus, Gyrostoma* 26
*Heliofungia actiniformis* 123
Heteractis
    *aurora* 19, 22, 28, 29, 30, 34, 36, 62, 64, 66, 70, 72, 74, 112, 116
    *crispa* 1, 18, 20, 21, 22, 28, **30**, 31, 34, 36, 62, *63*, 66, 72, 74, *75*, *76*, *77*, *82*, *83*, 86, 87, 90, 96, 98, 100, 102, 112, 116, 138, 142, *145*
    *magnifica* 2, *5*, *8*, *12*, *13*, *17*, *19*, 21, 30, **32**, 33, 42, 44, 45, *60*, *66*, *67*, *71*, 72, 74, 86, 90, 92, *93*, 94, *95*, 98, 100, 101, 116, *117*, *122*, *131*, *135*, *146*, *148*
    *malu* 20, 22, 28, **34**, *35*, 74
Hybrid Anemonefish 49, *85*, 133
*kenti*,
    *Stoichactis* 38, 40
*koseirensis*, *Radianthus* 28
*kuekenthali*, *Radianthus* 30
*latezonatus, Amphiprion* 31, 46, 59, **82**, *83*, 144
*latifasciatus, Amphiprion* 43, 56, 64, 66, 74, **84**, *85*, 96
Leathery Sea Anemone 30
*leucokranos, Amphiprion* 31, 33, 46, 49, 54, **86**, *87*, 100, *106*, *136*
*lofotensis, Urticina* 7
*Lutjanus*
    *ephippium* 76
    *percula* 98
*Macranthea cookei* 34
*Macrodactyla*
    *doreensis* 6, 20, 22, 28, 34, **36**, *37*, 70, 74, 100, 116, *117*
    *gelam* 36
*macrodactylus, Radianthus* 30
Madagascar Anemonefish 84
*magnifica*,
    *Actinia* 32
    *Heteractis* 2, *5*, *8*, *12*, *13*, *17*, *19*, 21, 30, **32**, 33, 42, 44, 45, *60*, *66*, *67*, *71*, 72, 74, 86, 90, 92, *93*, 94, *95*, 98, 100, 101, 116, *117*, *122*, *131*, *135*, *146*, *148*
Magnificent Sea Anemone 32

# INDEX

Maldives Anemonefish 92
*malu,*
   *Discosoma 34*
   *Heteractis* 20, 22, 28, **34**, *35,* 74
   *Radianthus* 30, 32, 36
Mauritian Anemonefish 70
*maxima, Parasicyonis 26*
McCulloch's Anemonefish 88
*mccullochi, Amphiprion* 27, 52, 53, **88**, *89,* 144
*melanopus, Amphiprion* 27, 31, 33, 46, *47,* 48, 49, 52, 53, 54, 76, 78, 88, **90**, 91, 120, 121, 125, 138
Merten's Sea Anemone 42
*mertensii, Stichodactyla* 2, *10, 17, 18,* 20, *21, 23,* 32, 38, 40, **42**, 43, 46, 60, 62, 64, 65, 70, 72, 73, 74, 80, 81, 84, *85, 87,* 94, 106, *107,* 112, 116, *119, 128, 136, 138, 142*
*nigripes, Amphiprion* 33, 46, 52, 53, **92**, *93, 100,* 110, *131, 135*
*nigrolineatus, Siokunichthys 123*
*ocellaris, Amphiprion* 33, 39, 43, 46, 58, **94**, *95,* 98, 125, *141,* 143, 144
Oman Anemonefish 96
*omanensis, Amphiprion* 27, 31, 56, 84, **96**, *97*
Orange Anemonefish 106
Orange-fin Anemonefish 72
*Oxylebias pictus 7*
*papillosa,*
   *Antheopsis 34*
   *Radianthus 34*
*Parasicyonis*
   *actinostoloides 26*
   *maxima 26*
*paumotensis, Radianthus 32*
*Perca polymnus 102*
*percula,*
   *Amphiprion* 10, 31, 33, *39, 45,* 46, 49, 58, 94, **98**, 99, *143,* 144, *145,* 148
   *Lutjanus* 98
*perideraion, Amphiprion* 2, *13, 17, 31, 33, 36, 39, 44,* 46, 54, 92, 100, 101, *110, 125, 132, 146*
*Physobrachia*
   *douglasi 26*
   *ramsayi 26*
Pink Anemonefish 100
*Oxylebias pictus 7*
*polymnus,*
   *Amphiprion 8, 31,* 41, 48, 55, **102**, 103, 108, *129,* 138, *139, 141, 142*
   *Perca* 102
Pomacentridae 45
*Premnas biaculeatus* 6, 11, 27, 48, 51, 59, **114**, 115, 131, 132, 138, 143
*Priapus giganteus 38*
*quadricolor,*
   *Entacmaea* 11, *18,* 19, 21, **26**, *27,* 62, *63,* 64, *65,* 66, 74, 76, *78, 79,* 88, *89, 90,* 91, *96, 97, 104, 105, 112,* 113, *114,* 115, *116,* 121, 122, 131, 137, 138, 139, 142
   *Gyrostoma* 26
*Radianthus*
   *gelam* 26, 36
   *koseirensis* 28

*kuekenthali 30*
*macrodactylus* 30, 32
*malu* 30, 32, 36
*papillosa 34*
*paumotensis 32*
*ritteri* 9, 30, 32
*simplex 28*
*ramsayi, Physobrachia 26*
Red and Black Anemonefish 90
Red Saddleback Anemonefish 76
*ritteri, Radianthus* 9, 30, 32
*rubrocinctus, Amphiprion* 27, 39, 48, 54, 76, 78, 90, **104**, *105*
Saddleback Anemonefish 102
*sandaracinos, Amphiprion* 43, 46, 53, 60, 86, 100, *106, 107*
Sea Anemone
   Adhesive 24
   Beaded 28
   Bulb-Tentacle 26
   Corkscrew-Tentacle 36
   Delicate 34
   Gigantic 38
   Haddon's 40
   Leathery 30
   Magnificent 32
   Merten's 42
*sebae, Amphiprion* 41, 48, 55, 102, **108**, *109, 141*
Sebae Anemonefish 108
Seychelles Anemonefish 80
*simplex, Radianthus 28*
*Siokunichthys nigrolineatus 123*
Skunk Anemonefish 60
Spine-cheek Anemonefish 114
*Stichodactyla*
   *gigantea* 9, 10, 20, 22, **38**, *39,* 40, 42, 66, 74, 94, 98, 99, *100,* 104, 116, *145*
   *haddoni* 8, 9, 10, 32, 38, **40**, 41, 42, 62, 70, 72, 74, 102, *103,* 108, 116, 138, *139,* 142
   *mertensii* 2, *10, 17,* 18, 20, 21, *23,* 32, 38, 40, **42**, 43, 46, 60, 62, 64, 65, 70, 72, 73, 74, 80, 81, 84, *85, 87,* 94, 106, *107,* 112, 116, *119, 128,* 133, 136, 138, 142
*Stoichactis*
   *digitata* 24
   *giganteum* 40, 42
   *haddoni* 40
   *kenti* 38, 40
*strasburgi, Dascyllus 116*
*Thalassoma amblycephalus 5*
*thiellei, Amphiprion* 54, 110, 111
Thiellei's Anemonefish 110
Three-band Anemonefish 112
Three-spot Dascyllus 116
Tomato Anemonefish 78
*tricinctus, Amphiprion* 27, 29, 31, 43, 48, 55, 59, 70, 72, 74, **112**, 113, *125, 141*
*trimaculatus, Dascyllus* 8, *37,* 45, 48, **116**, *117*
Two-band Anemonefish 66
*Urticina lofotensis 7*
White-bonnet Anemonefish 86
Wide-band Anemonefish 82

159

# NOTES